CHEMICALS IN THE ENVIRONMENT

Peeter Kruus
Marianne Demmer
Kierstan McCaw

CHEMICALS

IN THE

ENVIRONMENT

Peeter Kruus
Marianne Demmer
Kierstan McCaw

Published by

Polyscience Publications Inc.
Morin Heights, Quebec, Canada

ISBN 0-921317-26-3

Printed in Canada

Cover Photos Courtesy
North American Wetlands Conservation Council (Canada)

INTRODUCTION

Global change, urban smog, acid rain, polluted groundwater, heavy metal contamination - these are just a sampling of the major environmental problems we face today. They are examples of chemical pollution. They are the price that we are paying for what has often been a very beneficial reliance on chemicals in our industrial and personal lives.

We are bombarded with information about these and other pollution issues from across the spectrum of opinion. Despite this glut of information - or maybe because of it - most of us are left confused. While there is some relevant, authoritative, scientific evidence available about these issues, it is usually available only in "scientese", a language requiring a background in science for comprehension. The media fire bullets of information at us: the latest fire, accident or spill. Unfortunately, these reports are often one-dimensional accounts of an event or issue. Broader questions are rarely broached by the media: Where does this chemical come from? Why do we use it? Who benefits and how much do we rely on it? What other problems are connected with its use?

We wrote this book in response to what we felt was a need for information that sits somewhere between the experts and the media; written in language that everyone can understand, that provides some context for the major pollution issues and the chemicals involved.

Our intention was to act not as advocates or apologists, but as observers. We thus tried not to colour the issues with our personal opinions. However, at the end of the chapters we did allow ourselves to express some more personal views in sections titled "Concluding Remarks". As we explain in the last chapter on "Risk Assessment", the estimation of risk involves many uncertainties. It also involves some judgments of the values of consequences which are closely related to a person's life philosophy. The greatest problem with such risk estimation is the comparison of hazards to the health of individual humans and hazards to the ecosystem as a whole. The "Concluding Remarks" sections must be read with these uncertainties in mind.

We have put "descriptors" for each paragraph in the margins to try to make it easier for the readers to find their way through the discussions. There are no chemical formulas in the bulk of the text; these are given in an appendix.

For the sake of brevity and readability, the book does not contain the detailed references. Instead, some recommended readings are listed at the end of each chapter. Detailed references for the information can in most cases be found in these references or in the appropriate chapters in the second edition of "Controversial Chemicals" (P. Kruus and I.M. Valeriote, eds. Multiscience Publications, Montreal, 1984). A more detailed scientific presentation of many of the issues can be found in "Environmental Chemistry" (N.J. Bunce. Wuerz Publishing, Winnipeg, 1990).

Our aim has been to make this book interesting and easy to read for the general public. It is our hope that "Chemicals in the Environment" can also be of use in senior high school chemistry courses to make these more relevant to real problems in Canada and the world. The difficult part remains. We must as a society evaluate the multitudes of opinions and the volumes of information that surround us, and make decisions about how to address the problems arising from the presence of chemicals in the environment.

ACKNOWLEDGEMENTS

This book is built on two previous books prepared at Carleton University. The first book, "Controversial Chemicals: a Citizen's Guide", was published in 1979. It contained chapters on 25 chemicals which were of concern to the public at that time. A revised and expanded edition of "Controversial Chemicals" was published in 1984 by Multiscience Publications Limited, in conjunction with The Chemical Institute of Canada (CIC). This second edition contained 30 chapters describing the controversies underlying 30 chemicals or groups of chemicals.

"Chemicals in the Environment" is based to a significant degree on information available in "Controversial Chemicals". The range of chemicals covered has been limited to those involved in environmental issues, and the format of the book is quite different. Our work has been made much easier because of the efforts of those who assisted in preparing "Controversial Chemicals": Mary Valeriote, Mike Kazanel, David Sims, Tina Karel, Jim Holmes, Leanne Stuart, Jamie Donaldson, Paul Hope and Ruth Howard. We wish to acknowledge their contributions.

Many organizations and people have assisted us in preparation of this book. We have received financial assistance from the following organizations: Ontario Ministry of the Environment for support in the form of stipends for researchers for the first edition; Environment Canada for similar support for the second edition; The Canadian Chemical Producers' Association (CCPA) for support for stipends to K.M and M.D during writing of this book; CUSO and The International Organization of Consumers' Unions (IOCU) for support of P.K. during a sabbatical year at IOCU's office in Malaysia. Carleton University has contributed to this project through provision of space and other resources.

We are indebted to numerous people for assisting us by reviewing drafts of the book. Special thanks are due to Adrian Demayo, Bob Burk, Roger Pocklington, Keith Bourque and David Mowbray for their careful critiques. George Carmody gave us valuable help in setting up the wordprocessing system, and Maidie Armstrong and Lorena Duncan did a large amount of the wordprocessing. Annie Kruus and Robert Kehle assisted in the preparation of the figures. Special thanks are due to Mary Valeriote and Annie Kruus for editing and proof-reading.

None of these helpful people should be held responsible for any errors, misinterpretations or views expressed in this book. The authors are solely responsible for the content.

CONTENTS

CONTENTS

CONTENTS

CHAPTER I
OUTDOOR AIR POLLUTION

Smog alert in Toronto

Another day in Toronto's city centre; cars, trucks and buses clog the streets, people crowd the sidewalks. At 2:00 pm, radio stations announce a smog alert. Environment officials warn people with respiratory problems to stay indoors and urge everyone to avoid exercising outside. On this Thursday afternoon the 28th of October, 1982, the pollution index in the city reaches 51 - the highest smog level in seven years. The next day, the newspapers carry reports of many people suffering stinging eyes and increased difficulty in breathing. These are not very serious health effects, just as this was not a very severe smog.

Air pollution is a serious problem

Unfortunately, around the globe, incidents like the one described above are not at all uncommon. In the United States, some 150 million people breathe air considered to be unhealthy by the Environmental Protection Agency (EPA). In some urban areas in the US, jogging can be hazardous to your health because of air pollution. Hungarian health officials estimate that every twenty-fourth disability and every seventeenth death in that nation is caused by air pollution. The quality of the air is abysmal in large Third World cities such as Mexico City, Beijing, Calcutta, Sao Paulo, Jakarta and Tehran. At the same time, acid rain caused by air pollution is acidifying lakes and damaging forests. These are only some examples of outdoor air pollution.

Pollution of water and soil through air

We emit large amounts of a wide variety of pollutants into our atmosphere. Some of these can harm people and the environment, not while they are in the air, but after they have been deposited. As much as 90% of the toxic chemicals that enter Lake Superior come from the atmosphere carried by rain and snow; the problems caused by some of these pollutants are discussed in the following Chapters. The hazards of municipal waste incinerators seem to arise not only from breathing air polluted by emissions from their stacks, but from eating food contaminated by chemicals such as cadmium (see Chapter VII) and chlorinated furans (see Chapter V) which are deposited down-wind.

PAHs are covered in Chapter III

Some chemicals are air pollutants both outdoors and indoors; polyaromatic hydrocarbons (PAHs) come from forest fires, burning oil wells, wood stoves, truck exhausts and cigarette smoke. PAHs are a serious problem in indoor air pollution, and are discussed in Chapter III. The "tire fire" at Hagersville Ontario and the burning oil fields in Kuwait will only cause temporary pollution of the atmosphere by substances such as PAHs. The pollution of soil and water will be more serious and persistent.

Chapter I deals with smog and acid rain

Two primary types of outdoor air pollution are discussed in this chapter: smog and acid rain. Both of these are complex problems; indeed it seems that the more research done on them, the more complex they become. Basic to an understanding of these air pollution problems, though, is an understanding of the dynamics of the air itself.

THE AIR NATURE PROVIDES US

Dynamic nature of the atmosphere

Our atmosphere is a dynamic system. Components are continuously being added and removed; yet the final outcome is an atmosphere with a remarkably constant composition. This atmosphere is not very representative of the cloud of gas and dust from which the earth formed. Instead, it reflects the complex interactions between those primordial dust clouds and the rocks and oceans of the earth's crust together with the earliest living things themselves. Our atmosphere is dependent not only on geological and chemical processes, but also on the biological processes that occur on earth.

Composition at sea level

The composition of the atmosphere is not constant. In this chapter, we will be discussing air near sea level, in the troposphere. In Chapter II, we will also be concerned with the composition of the air at higher altitudes, in the stratosphere. Nitrogen (N_2), oxygen (O_2), and argon (Ar) are the major chemical constituents of dry air at sea level; they account for roughly 79, 20 and one percent, respectively.

Importance of nitrogen to oxygen ratio

Nitrogen is not very active biogeochemically and has thus a very long lifetime in the atmosphere, of the order of millions of years. Oxygen cycles through the atmosphere, oceans, biosphere and sedimentary rocks; it has an atmospheric lifetime of the order of thousands of years. The proportion of oxygen to nitrogen is very important for maintaining life on this planet. Oxygen is essential for life, but nitrogen is also important since oxygen must be diluted. If oxygen levels were to rise from 20% to 30%, devastating fires caused by lightning would probably destroy all vegetation, while less than 15% oxygen would make it impossible to start any fires.

Argon is inert

Most of the argon in the atmosphere has been released into the air through volcanic activity and is produced by the radioactive decay of an isotope of potassium in the earth (see Appendix B). Argon is biogeochemically inert; once released into the air, it stays there.

Water acts as a thermostat

Water vapour is another important component of the atmosphere that occurs in amounts which vary widely, ranging from four percent by volume in warm humid weather to under one percent. It is the cycling of water through the atmosphere that produces clouds, rain and snow, and that cools the biosphere through evaporation. Water vapour is also an important "greenhouse gas", keeping the climate of our planet hospitable to life.

Very little of other gases in the atmosphere

The remaining trace gases which occur in air are of such small quantities that their concentrations are usually given in parts per million (ppm) rather than per hundred (percent). A very important greenhouse gas, carbon dioxide (CO_2), is the most abundant of these at roughly 340 ppm. The next most common are the inert gases, neon at 18 ppm and helium at five ppm.

Ozone concentrations in the atmosphere vary with altitude, reaching a maximum of 12 ppm in the ozone region of the stratosphere (about 30 kilometres up). There, ozone plays an important role in shielding the earth's surface from ultraviolet radiation (see Chapter II). The concentration of ozone near the ground is usually quite small, but it becomes greater when photochemical smog is present.

Ozone is more common in the stratosphere

The atmosphere recycles other chemicals essential to life or to the maintenance of a stable climate. For example, the dimethyl sulphide (DMS) produced by plankton in the oceans plays an important role in rain formation. Apparently, the DMS produced by the plankton is oxidized in the atmosphere leaving droplets of sulphuric acid which can dissolve in atmospheric water vapour, and produce clouds which may subsequently cause rain. This is essentially the same process that leads to the formation of acid rain from industrial emissions of sulphur compounds.

Chemicals are recycled by the atmosphere

The air has some ability to cleanse itself. Water vapour in the atmosphere condenses on "nucleation centres" such as the droplets described above, tiny particles of dust, or salt from sea-spray. Various pollutants in the atmosphere are then dissolved in these snow-flakes or rain-drops as they grow and fall. The dissolution of carbon dioxide from the atmosphere into rain means that natural rain is actually weakly acidic, with a pH of about 5.6.

Precipitation cleanses the atmosphere

The dynamic nature of air is one reason why air pollution control is difficult. The vast majority of chemicals emitted to the atmosphere react rapidly and often unpredictably. For example, the ozone present in dangerous urban smog is not actually emitted by human activities in significant amounts. Rather, various hydrocarbons (often referred to as VOCs, "volatile organic compounds") and nitrogen oxides (called NOx) are released to the air, mainly by motor vehicles. These undergo reactions with oxygen in the presence of sunlight, forming ozone as a by-product.

Chemical dis-equilibrium of our atmosphere

The earth's atmosphere is very different from those of its neighbours Mars and Venus. Their atmospheres are largely composed of carbon dioxide and exist in a state of chemical equilibrium with the surfaces of these planets. However, the earth's atmosphere, although in a steady state, is actually in chemical non-equilibrium. Without life on earth, most of the oxygen in the atmosphere would disappear. There is 20% oxygen in the atmosphere only because it is formed in the fundamental biological process of photosynthesis. Our planet's atmosphere has been maintained in such a non-equilibrium system for millions of years only due to the presence of life on earth.

Atmospheres of Mars, Venus in equilibrium

In the 1970s, the "Gaia hypothesis" was proposed to explain the earth's unique atmospheric composition. (Gaia is the name of the Greek goddess of the earth.) Lovelock suggests that "the dynamic forces of life so dominate our planet that life has a controlling influence on the oceans and atmosphere". In other words, in some way, the presence of life itself allows the creation and perpetuation of an atmosphere conducive to life.

Atmospheric composition depends on life

URBAN AIR POLLUTION

Industrial revolution and air pollution

The industrial revolution brought about tremendous changes in technology, society and the environment. Among these was the advent of air pollution as a serious threat to human health. A century ago, many cities in Europe and the United States were covered in shrouds of sooty smoke from the burning of coal. Days where the air was still and heavy resulted in sickness and even death. In 1952, the infamous "black fog" engulfed London, England, for several days. About 5000 people died and tens of thousands became ill.

Many cities have serious air pollution

The problem of air pollution in the centres of many cities is still a critical issue. In Bombay, India, a disaster of the magnitude of the tragedy in London is a real possibility. Although industrial emissions and combustion of wood and waste by an increasing population of slum-dwellers contributes to the problem, the pollution in Bombay is primarily due to a rapid increase in the number of motor vehicles. The number of cars on the roads in Bombay has doubled in the past decade. Most of these cars are old and badly maintained, and there are essentially no emission control laws. Delhi in India may have an even worse problem. It has over one million motor vehicles emitting 250 tonnes of pollutants every day into a geographic setting where there are many periods without heavy rains or winds to cleanse the air. Many other cities are becoming infamous for the poor quality of their air: Los Angeles, Santiago, Athens, Tehran, Mexico City, Beijing and Bangkok are prime examples.

Air pollution in Eastern Europe

The air pollution problem in parts of Eastern Europe is also quite serious. This is due largely to their reliance on high-sulphur brown coal as an energy source, which results in large emissions of soot and oxides of sulphur and nitrogen. Sulphur emissions from East European factories, power plants, and residences totalled 17 million tonnes in 1989. The town of Bitterfeld in what was East Germany has often been referred to as the most polluted place on earth; statistics show that, on the average, the residents of Bitterfeld die five to eight years earlier than the national average. Parts of Czechoslovakia, Poland and Romania have similar problems.

Smog a major problem in North America

In North America, the major problem in urban air pollution is photochemical smog, in particular the ozone it contains. The 11 million residents of the Los Angeles area were exposed to 172 "unhealthy" days in 1989, with ozone levels reaching three times the federal health standard. The brown, hazy smog decreases visibility, makes people's eyes sting and often has an unpleasant, bleach-like odour. There is growing evidence that urban air pollution is damaging the lungs of healthy people, making them more susceptible to respiratory diseases like emphysema and chronic bronchitis.

Smog also affects plants

Plants are also affected by smog. In Ontario, a 1984 Environment Ministry report concluded that more than a dozen cash crops in the province had been damaged by high ozone levels. The affected crops include white beans, potatoes, grapes, tobacco, tomatoes and onions. Apparently, the ozone damages the leaves of plants causing premature yellowing and spotting which can stunt plant growth and reduce yields.

Nitrogen Oxides

Formation of nitrogen oxides

Nitrogen oxides (NOx) play an important role in the formation of smog. They are formed naturally as well as through industrial processes. Nearly all of the nitrogen in the air is in the form N_2, but some of it exists in the form of three nitrogen oxides: nitrous oxide (N_2O), nitric oxide (NO) and nitrogen dioxide (NO_2). Some is also present in the form of ammonia (NH_3). These compounds enter the atmosphere from various sources. Soil bacteria decompose dead biological matter, with the resulting nitrogen compounds either being taken up by the surrounding vegetation or emitted to the air. Nitrogen oxides are formed wherever air becomes very hot: in lightning bolts, engines, cooking flames and cigarettes. The combustion of nitrogen-containing materials such as some coals also releases NOx into the air. Ammonia is also produced industrially from N_2 for use primarily as fertilizer; this subsequently enters the "nitrogen cycle".

Major steps in the nitrogen cycle

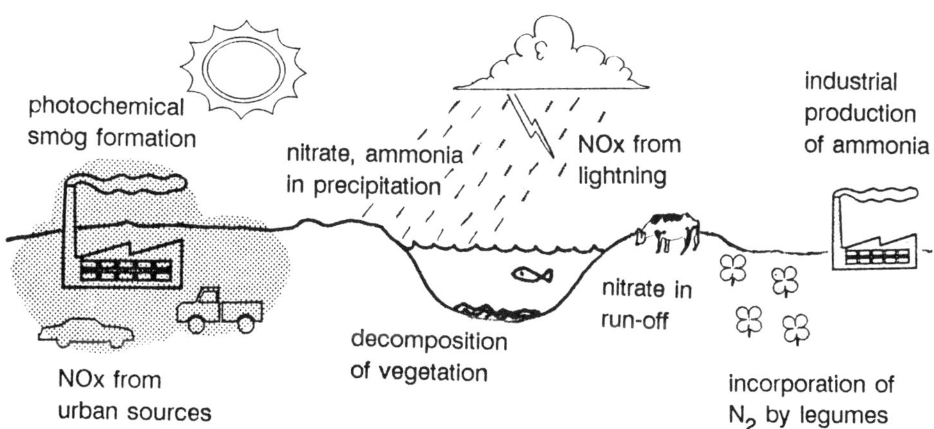

Mechanisms of nitrogen removal

Nitrogen is removed continuously from the atmosphere in various ways in the nitrogen cycle, resulting in a steady state concentration of NOx and ammonia. Nitrogen oxides are removed by biological processes, or through reactions in the air to form nitric acid (HNO_3) which dissolves in water and is thus cleansed from the air through precipitation. Ammonia can return to the soil or water directly from its gaseous state. It can also come down in precipitation, usually in the form of a nitrate or a sulphate salt. In this way, the presence of ammonia in the atmosphere is important in moderating the acidity of rain.

Our activities influence the natural cycle

Our industrial society is now significantly influencing the natural balancing processes which maintain the levels of various nitrogen compounds in the atmosphere. Even though the NOx introduced by human activities are only a small fraction of the total atmospheric production, they have significant effects because they are concentrated in urban areas.

Motor vehicles emit NOx

Gasoline powered vehicles are the largest source of NOx emissions, but other important sources include diesel engines, industrial fuel combustion, power plants and forest fires. Approximately 2 million tonnes of nitrogen oxides per year were emitted in Canada in the mid 1980s; about ten times that amount were emitted in the US. In Canada, roughly two-thirds of the total NOx emissions came from the internal combustion engines of transportation vehicles. The remainder were from various industrial, commercial and residential activities.

Nitrous oxide (N_2O) a greenhouse gas

Increased levels of NOx in the atmosphere can have effects on the environment and human health. Nitrous oxide, (N_2O) otherwise known of as "laughing gas" is not known to be toxic to humans and is familiar to some people as an anaesthetic. Nitrous oxide does, however, contribute to the greenhouse effect (see Chapter II).

Nitric oxide (NO) from fossil fuels

The majority of the nitrogen oxides emitted by fossil fuel combustion are in the form of nitric oxide (NO). In the atmosphere, nitric oxide is converted to nitrogen dioxide in a complex sequence of reactions. This process is usually rapid, but may take hours or days if the effluent gases from a flame are cooled quickly.

Nitrogen dioxide (NO_2) is toxic

Nitrogen dioxide is the most toxic of the nitrogen oxides. Exposure of humans to 150 ppm or greater of nitrogen dioxide can cause death from pulmonary edema. Lower levels of exposure lead to an increase in the resistance of air passages in lungs, but the effects seem to be reversible. Nitrogen dioxide also causes injury to vegetation and fading of dyes.

Carcinogenicity of NOx

Nitrogen oxides have not been demonstrated to be carcinogenic. Yet a correlation has been noted between incidence of cancer and high levels of NOx. It is possible that the cancers may be caused by chemicals which are formed by reactions between nitrous acid derived from NOx and amines like ammonia. These are called nitrosamines. One of these, dimethylnitrosamine (DMN), is a proven potent carcinogen in animals. Nitrosamines have been detected in places where traffic is the only apparent source. It has, however, not been demonstrated conclusively that nitrosamines are formed in our bodies following inhalation of nitrogen dioxide.

Smog

By-products of NOx serious pollutants

The presence of nitrogen oxides in urban air is a serious problem, but the hazards posed by chemicals formed through reactions involving NOx are far more severe. The most important of these chemicals is ozone, which plays a much larger part in the health and environmental impacts of smog than nitrogen oxides, and appears to be involved in die-back of forests.

Ozone is a dangerous pollutant

Ground level ozone is formed naturally in small quantities in electric discharges in the air, such as lightning, and gives the air that characteristic "fresh" smell. Larger amounts of ozone near ground level can, however, be quite harmful. Ozone has the same type of toxic effects as nitrogen dioxide, but to a considerably greater degree. Both ozone and nitrogen dioxide can oxidize biological compounds such as fatty acids. According to the United States Centre for Disease Control, ground level ozone can weaken the immune system and destroys lung tissue. A day spent breathing air with an ozone level of 0.14 ppm can harm active people like joggers, construction workers and children playing outside. If we are exposed for only one hour to 0.08 ppm of ozone (the Canadian standard), the airways in our lungs become inflamed.

Unfortunately, the task of reducing ozone levels in our cities is proving to be very complicated. Ozone levels still regularly exceed desirable levels in many cities in Canada and US.

Reducing ozone levels is difficult

Ozone in urban air seems to be formed in the "photolysis" of nitrogen dioxide. During this "photochemical" process, nitrogen dioxide absorbs sunlight, causing the release of a single oxygen atom (O). The oxygen atom is then free to combine with molecular oxygen (O_2) to give ozone (O_3).

Source of ground level ozone

The reactions which create and break down ozone give rise to an interesting phenomenon. In the immediate vicinity of NOx emissions, there is a predominance of nitric oxide (NO). It breaks down ozone, causing a decrease in the quantities of that toxic molecule. However, in regions further downwind from the NOx source, most of the nitric oxide has been converted to nitrogen dioxide, which reacts to increase the ozone level.

More ozone downwind from NOx source

The ozone that is formed does not last forever. It will not accumulate in the air if it can react with NO to re-form NO_2 and molecular oxygen. But many other reactions are possible in the mixture of chemicals typical of an industrialized urban area. Increased emissions of NOx are usually associated with emissions of VOCs, primarily hydrocarbons. Almost half of the hydrocarbons in the air are produced by motor vehicles and consist largely of unburned gasoline.

VOCs add to the pollution problem

The presence of VOCs makes many other photochemical reactions possible. Some of the products can be strong eye irritants, such as peroxyacetylnitrate (PAN). It is also extremely toxic to plants; damage usually appears as a silvery or bronzed coloration on the lower leaf surface. Recent reports have found traces of PAN over the Pacific Ocean, indicating that PAN can be present even in clean air.

NOx plus VOCs plus sunlight gives smog

Regulations have been introduced in some regions to try to reduce the hydrocarbons emitted from vehicles, but other sources of VOCs are more difficult to control. Organic chemicals evaporate from hundreds of different sources, ranging from large factories to paint to home cleaning fluids.

Sources of VOCs in urban areas

Trees and other vegetation also naturally emit VOCs, but their contribution to pollution in urban centres is small in comparison to that from human sources. Natural VOCs have thus in the past been thought to be insignificant in ozone formation. However, recent research indicates that peak ozone concentrations occur about 50 kilometres downwind of urban centres. There, the levels of biologically produced VOCs may approach those experienced in cities, especially if the area is heavily forested. Smog in such an area can harm the vegetation.

Natural sources of VOCs

Controlling Smog

It is difficult to control both NOx and hydrocarbon emissions from engines. Conditions like rich mixtures of fuel, which produce little NOx, tend to produce considerable quantities of hydrocarbons. There is less air pollution overall if lean

Catalytic converters reduce smog

fuel mixtures (mixtures involving a considerable excess of air) are used, but most engines do not perform well on lean mixtures. One way to reduce both NOx and hydrocarbons in automobile emissions involves the use of a catalytic converter. Such converters require the use of unleaded gasoline, since lead deactivates the catalyst found to be most effective in the converter.

Smog problem in Los Angeles persists

Because of the serious smog problems in Los Angeles, the State of California imposed the first automobile emission standards in 1966. These have been made more stringent through the years. Now less pollution is produced by the over 8 million vehicles there than from the 2 million that were on the roads in the early 1950s. But the smog problem persists. New standards have been announced; new cars sold by the year 2000 will be required to emit less than 20% of the pollutants allowed in emissions today. This may require the introduction of cleaner "reformulated" gasolines, a switch to another fuel such as methanol, propane or natural gas, or to large scale use of electric vehicles.

Clean Air Act means many changes

In January of 1989, the California Clean Air Act went into effect. The people of Southern California may have to make dramatic changes in their lifestyle to meet the demands of this legislation. The Act requires a statewide reduction of smog-creating pollutants by five percent per year starting in 1990. In the long term, the solution may involve restructuring of urban areas and lifestyles to lessen the dependence on automobiles.

Lifestyle changes may be needed

Air pollution from vehicle emissions will be reduced by changes in people's habits in addition to changes in the vehicles and the fuels. Several long-range schemes being planned for the Los Angeles area focus on reducing the need to drive, for example phasing in work-weeks of four ten-hour days wherever possible, and restructuring suburban neighbourhoods so that employment spaces and housing are closer together in "neighbourhood work stations".

Many activities will be affected

In order to lower the emissions of VOCs, there will be a ban on volatile fluids for starting barbecues and on gasoline powered lawn-mowers. Bakers will have to control ethanol emissions from yeast and farmers will have to dispose of manure in acceptable ways to lower emissions of nitrogen compounds. But such changes will not come cheaply. Implementing the measures will probably cost between $600-$2200 per home per year.

Third World countries have more problems

As the environmental and health impacts of air pollution become increasingly well understood, the pressures mount for governments to take more effective measures to control emissions of the polluting chemicals. It is easier to introduce such legislation in states like California, where the problem is obvious, and where there are resources available to pay for the higher cost of stricter emission standards. Air pollution from automotive exhausts may be just as bad in cities in less industrially developed countries (Third World countries). The imposition of more stringent and more costly standards is more difficult there, as people do not have the resources to upgrade their vehicles or to use more expensive fuel.

The world will be watching what happens in Southern California for the next twenty years. It is the place most likely to lead the way in urban air pollution control.

Los Angeles might lead the way to clean air

ACID RAIN

We became aware of the problem of "acid rain" in the late 1960s. Scandinavian scientists, seeking causes for the declining fish stocks in their lakes, began to suspect sulphur dioxide emissions from some of their industrialized neighbours in Europe. By the early 1980s, there were signs of widespread damage to the forests of West Germany. Scientists placed the blame on acid deposition from rain and snow, as well as dry deposition from dust. The percentage of forests in Germany showing signs of damage rose dramatically in the six years from 1982 to 1988, from eight to 52%.

Acid rain noticed first in 1960s

Acid rain, more properly acidic precipitation, is usually defined as precipitation having an acidity below pH 5.6. "Clean rain" is naturally slightly acidic with a pH of about 5.6, due to the presence of carbon dioxide in the atmosphere. But some of the rain falling on parts of eastern Canada is up to 30 times more acidic than "natural" rain.

Definition of acid rain

Acidic — Basic

0 1 2 3 4 5 6 7 8 9 10 11 12 13 14

stomach acids (1–2); vinegar, lemon juice, Coke (2–3); clean water (5–6); milk (6–7); baking soda (8–9); detergents (10–11); Draño, machine dishwashing detergent (12–14)

pH Scale

Neutral pH is pH 7

The pH scale measures acidity. At a pH of 0, a solution is very acidic (it has a high concentration of hydrogen ions), while at pH 14, it is very alkaline (high in hydroxyl ions, low in hydrogen ions). A pH of 7 represents a neutral solution where the acidic and alkaline components are in equal concentrations, and thus neutralize each other. The pH scale is logarithmic, so that solutions that differ by one pH unit have a tenfold difference in hydrogen ion concentrations. Some perspective can be gained by realizing that vinegar has a pH of less than 3, milk has a pH of about 6.6, and the pH of a baking soda solution is just over 8.

pH is a measure of acidity (hydrogen ions)

Acid rain increases the rate of corrosion and weathering of materials. It contributes substantially to human health problems, and is an ecological hazard. Over 150,000 lakes in eastern Canada are believed to be suffering biological damage, and 19 salmon rivers can no longer support their species. In addition, more than 50% of the forests in eastern Canada grow in areas where the rainfall is unusually acidic.

Damage caused by acid rain in Canada

US and Europe are also affected

In the northeastern United States, damage has been evident for some time. Recently, several of the southeastern states, much of the Great Lakes area and the western mountainous regions also appear to be at risk. In Europe, the list of vulnerable areas in different nations is growing, although Sweden and Norway are still thought to have experienced the most severe damage to their lakes.

Sources

Sulphur and nitrogen oxides cause acid rain

Acid rain is due primarily to emissions of sulphur oxides into the atmosphere where they can form sulphuric acid. However, nitrogen oxides in the air can form nitric acid which can also acidify rain. In France, it is estimated that the contribution of nitrogen oxides to acid rain is up to 46%. In Britain, about 30% of acid rain comes from NOx.

The problem is international

The sulphur and nitrogen oxides in the atmosphere can dissolve in atmospheric water and then travel up to thousands of kilometres with the prevailing winds, eventually coming down as rain or snow. While urban air pollution is a more local problem, acid rain is often a problem on an international scale. Since acid rain can fall far downwind of the point of original emissions, the United Kingdom has often been blamed for the acid rain which falls on the lakes and rivers of Norway and Sweden. It now appears that sources in Poland and East Germany are more at fault. Acid rain from the US reaches Canada, and vice versa. A joint government-sponsored Canada-US research group estimated that about three to four times as much sulphur moves from the US to Canada than the reverse, partly due to prevailing wind patterns.

Sources are difficult to identify

The mobility of airborne pollutants makes it difficult to identify specific sources of acid rain. Labelling the sources is even more problematic because areas experiencing rain with pH below 5.6 have been observed many thousands of kilometres from industrial activity. The extra acidity may be due to sulphuric acid from volcanic eruptions or nitric acid produced when nitrogen oxides that are formed in lightning storms dissolve in rain.

Sulphur dioxide is the major cause

The major culprit in the acid rain problem, sulphur dioxide (SO_2), is a colourless gas which has a pungent and irritating odour at concentrations above three ppm. Sulphur dioxide is the primary oxide of sulphur emitted into the atmosphere through industrial activities, although sulphur trioxide (SO_3) can contribute up to ten percent of total sulphur emission. The combustion of any sulphur-containing fuel involves a reaction between sulphur and oxygen to produce SO_2. It can then react further to produce SO_3; this reaction is made more rapid (catalyzed) by the presence of hydrocarbons or particulates. In turn, the SO_3 dissolves in atmospheric water, forming sulphuric acid (H_2SO_4) thus increasing the acidity of the water which will eventually fall as precipitation to the ground.

We emit large amounts of SO_2 into air

The amount of SO_2 entering the global atmosphere from human activities is now of the same magnitude as natural emissions. The burning of coal accounts for about half of all industrial SO_2 emissions, with metallurgy, transportation, solid waste disposal and other industrial processes contributing the rest. Unlike natural SO_2 emissions, our emissions tend to be highly concentrated in areas of heavy urban and industrial activity.

Natural sources of SO_2

Sulphur dioxide is produced directly in nature from forest fires and volcanoes. When Mount St. Helens erupted in 1980, it was found that 90% of the aerosol mass it spewed out was composed of sulphuric acid droplets. The eruption of El Chicon in 1982 also introduced large amounts of SO_2 into the atmosphere. Sulphur dioxide is also formed in the oxidation of hydrogen sulphide (H_2S), which arises primarily from biological decay processes such as those occurring in swamps, tidal flats and estuaries.

Neutralizing acid gases in the atmosphere

Neutralization of acidic compounds can occur naturally in the atmosphere; ammonia is one substance which can neutralize the acidity of nitrogen and sulphur oxides. The rain in northern China is often alkaline (with a pH greater than 7) even though large amounts of SO_2 are emitted there. This is because the soil there is generally alkaline and the air contains high concentrations of soil dust, especially from deserts.

Rates of transport are in millions of tonnes per year

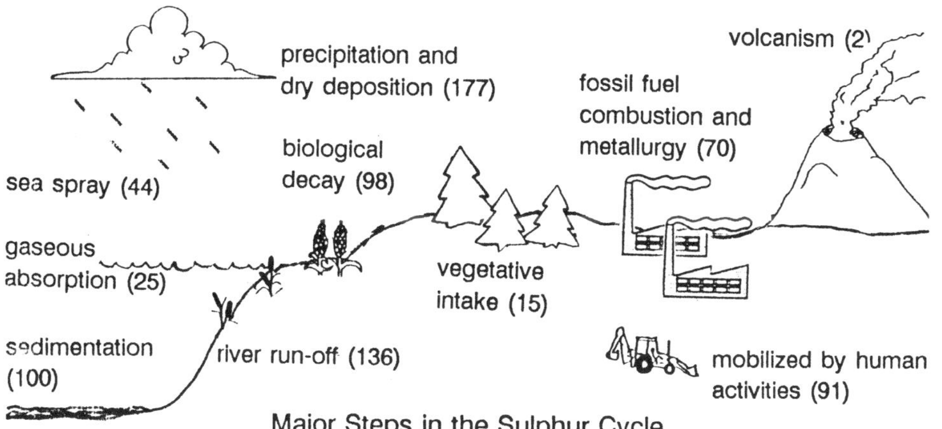

Major Steps in the Sulphur Cycle

Emission of SO_2 in Canada has decreased

Over five million tonnes of SO_2 were emitted in Eastern Canada in 1970 through human activity. The main sources were non-ferrous smelters (primarily nickel and copper), coal-burning electrical generating stations, petroleum refining, natural gas processing, transport and solid waste disposal. By 1980, these emissions had been reduced to under four million tonnes and by 1986 to less than three million tonnes. These reductions are continuing as new technology is introduced, especially in the metallurgical industry and in electricity generation.

Emissions from sources in USA

In the 1970s, nearly 30 million tonnes of SO_2 were emitted annually in the US from human sources, with electrical utilities accounting for 65% of that total. In the upper Ohio Valley (eastern Ohio, northern West Virginia, and western Pennsylvania) many large power plants burned high sulphur coal with little control over their sulphur emissions. Other industrial sources accounted for nearly 25% of the total, while nonferrous smelters emitted another 10%. By 1987, the U.S. had cut their annual emissions by about 25%. These reductions will continue as the new Clean Air Law which was passed in the US in 1990 will be taking effect.

Effects

SO_2 is a respiratory irritant

High concentrations of SO_2 in urban areas contribute to smog formation in city centres. Sulphur dioxide is one of the many pollutants that irritate the respiratory tract. The World Health Organization (WHO) suggests that effects would begin to appear at the following levels of SO_2: 0.04 ppm respiratory symptoms; 0.11 ppm, worsening of patients with pulmonary diseases; and at 0.19 ppm, excess deaths and illnesses. People with asthma, the elderly or others susceptible to chronic respiratory problems may be affected by concentrations lower than these.

Acid rain also affects human health

Sulphur dioxide is also a potential health hazard when it reaches us in the form of acid rain. Acid rain can cause the leaching of toxic chemicals from watersheds into rivers, lakes and groundwater, leading to the contamination of drinking water. Studies in Sweden and North America indicate that as the acidity of lake water increases, so does the level of methylmercury found in fish. This form of mercury, which is much more toxic than metallic mercury, is more common in acidic water, and is easily ingested by fish and other aquatic organisms (see Chapter VII).

Acid rain attacks materials

Acid rain also attacks metals, and structural materials such as limestone and marble. The cost of accelerated decay of structures such as bridges by acid rain is considerable. It is estimated to be hundreds of millions of dollars in Canada.

Limestone neutralizes acidity in lakes

Lakes and streams can become too acidic to support certain aquatic life. This is especially true in areas which do not have much material, such as limestone, that can neutralize the acid rain as it runs off into the body of water. The Canadian Shield is one such area. In other areas such as the Prairies, lakes and streams are often naturally alkaline, ie they have a pH greater than 7, and can thus neutralize the acidic precipitation. In the past decade, some lakes in Ontario that were severely damaged by acidification have started to recover as SO_2 emissions and hence acidic precipitation has decreased.

Acid rain may be a cause of forest die-back

There are indications that acid rain is increasing the acidity of forest soils. Damage to forests has been particularly evident in Germany. Many explanations have been offered for the "forest die-back" that is occurring. Besides acidification of the soil, variations in rainfall and temperatures have been suggested as causes. The combined effects of several pollutants such as acid rain, metals and ozone may make forests susceptible to many natural stresses such as dry periods, early frost and diseases.

Ozone may be responsible for forest damage

Although SO_2 is still thought to be the major reason for the acidification of lakes, there is increasing evidence that ground level ozone is the real culprit in the mysterious deaths of forests around the world. Experiments conducted on the trees and soils of affected areas showed both to be apparently unaffected by acidity equal to that found in the rain. However, after ozone had been added to their environment, the trees began to display signs of damage.

Measurements of ozone in declining forests found that it was present in the air at levels shown to cripple trees. This might help to explain why many forests started dying even after SO_2 emissions had been lowered for several years.

Economic costs of acid rain

The economic costs of acid rain can only be roughly estimated. In northern Ontario, the local fishing industry forms the basis of the area's tourist industry. It has been estimated that the total annual loss to this area due to acid rain could reach $230 million. In Nova Scotia, nine rivers have become too acidic to support salmon, resulting in an estimated $300 million annual loss to the local economy. The annual damage caused by acid rain to building materials in Canada is estimated to be at least $285 million.

Regulations

Debates over the effects of acid rain

A serious impediment to dealing with acid rain pollution has been that much of the research on its environmental effects has not been definitive. The Edison Electric Institute, an association of American electric companies, has said that there is "little evidence that variations in acidity have had adverse impacts on the environment". More recently, the National Acid Precipitation Assessment Program set up by the US Congress to analyze the impacts of acid rain, has reported that "acid rain is a problem, but it is not the extreme problem that is sometimes reported...".

Views from USA and Canada

Up to a few years ago, the US government's position on the matter was that there was too much scientific uncertainty about the sources and effects of acid rain to set pollution limits. The Canadian parliamentary subcommittee on acid rain has recognized that "the data base has to be substantially increased before firm conclusions can be reached about the effect acid rain is having...". But in the words of a member of the Canadian Forestry Service: "It is not... (appropriate that)...a patient should die before being considered ill".

Federal legislation in Canada

Regulation of pollutants such as SO_2 in Canada is complex. The federal government deals with air pollution primarily through its Clean Air Act. The Act sets ambient air objectives for the quantity and concentration of specific pollutants. For example, the federal government has set a "tolerable" range for SO_2 of 300 to 800 micrograms per cubic metre over a 24-hour period, and has adopted emission guidelines for new thermal power plants.

Involvement of provincial governments

The federal government shares jurisdiction over environmental matters with the provinces. Many standards and guidelines provided in the Clean Air Act require the approval of, and are enforced by, the individual provincial governments. Provincial consent is not required when limits are set on emissions from stationary sources that are a threat to human health or that would cause Canada to violate any international obligation it has undertaken with reference to air pollution.

Canadian plan to reduce acid rain

Since 1988, the federal and provincial governments of Canada have been co-operating to reduce emissions causing acid rain. Each province has implemented an abatement program designed to achieve national reduction targets. The aims of this program, which will cost the federal government roughly $330 million, include: a 50% reduction in SO_2 emissions by 1994; new standards which will reduce allowable emissions of NOx from cars and light-duty trucks by 67% and emissions from heavy-duty trucks by 50%; further research into the effects of acid rain; development of pollution control technology; and increased use of low sulphur coal by utilities.

Clean Air Act in USA will reduce acid rain

For the past few years, the Canadian government has been lobbying the American government in an effort to begin a concerted fight against acid rain. Finally, at the end of 1990, the US passed a new Clean Air Act. Emissions contributing to acid rain will be reduced, since electric utilities face strict cutbacks in SO_2 and NOx emissions. Automobile makers will have to design vehicles with reduced emissions from both the fuel and the tailpipes. Industries, municipalities and individuals will have to reduce emissions such as VOCs, PAHs and other hazardous air pollutants.

Cost of the Clean Air Act

By the time the requirements are in full force in 2005, compliance is estimated to cost about $25 billion per year. Consumers will have to face the rising costs of services and products affected by the bill, estimated at about $20 per month per household by 2005.

Control needs international agreement

Air pollution problems such as acid rain do not respect national boundaries. In 1991, Canada, US and 23 European governments signed the first major international cross-border pollution agreement. In March 1991, Canada and the US also signed a bilateral agreement to control emissions causing acid rain. It may be that an International Law of the Atmosphere can be agreed upon at the United Nations Conference on Environment and Development that will be held in Brazil in 1992. If we are to solve problems such as acid rain, and more importantly the problems discussed in Chapter II, then we will require cooperation of the majority of the governments of the world.

Technology

Methods for reducing SO_2

A considerable proportion of SO_2 pollution is the result of the combustion of fossil fuels. Reduction of acid rain is thus related to lowering the greenhouse effect (see Chapter II). Methods for reducing SO_2 emissions from fuel combustion include: energy conservation, resulting in reduced fuel consumption; a move to fuels with lower sulphur content; substitution of fossil fuels by other energy sources such as nuclear, solar, wind or renewable sources; and reduction of the sulphur content of fuel before combustion (coal washing, coal gasification, desulphurization of liquid fuels). Alternatively, it may be possible to reduce the formation of SO_2 during combustion (fluidized bed combustion) or reduce the emission of sulphur dioxide after combustion (stack gas desulphurization).

In non-ferrous smelting, sulphur dioxide emissions can be reduced by removing some sulphur from the ore before roasting. It is also possible to recover sulphur dioxide from flue gases, mainly in the form of sulphuric acid, which can be marketed. This has already been done by some companies for over 50 years.

Recovery of SO_2 from smelters

Attempts are also being made to mitigate the effects of acid rain. Liming programs, in which limestone or other alkaline substances are added to lakes, are used to neutralize the acid rain and increase the lakes' pH.

Attempts to mitigate effects

CONCLUDING REMARKS

Outdoor air pollution poses a serious risk to human health in many regions. In the Western world, people in some large urban areas are at risk. Although outdoor air pollution from industrial emissions has now decreased to a considerable extent in some countries, the problem of urban smog persists. In Eastern Europe, pollution from heavy industry creates health problems in several regions. In Third World countries, both urban smog and industrial pollution are serious health risks for people living there.

Air pollution is a serious health hazard

There is a risk to the environment from acid rain, ozone and other air pollutants. The ill-effects are serious in some regions, for example acidification of lakes in Scandinavia, Quebec, New York State and Ontario, and forest die-back in Germany. However it still seems possible to reverse these situations within a few decades.

Acid rain threatens the environment

The outdoor air pollution problems discussed here are regional. The pollutants do not remain in the atmosphere for such a long time that they present global problems, in contrast to the effects discussed in the next chapter. Problems such as acid rain, and even urban smog, are nevertheless often international in scope.

The problems are inter-national

There is improvement in some regions. In Canada, the acid rain problem has been diminishing in the last decade; some acidified lakes are starting to recover. Particulate emissions (dirt, dust, soot) have decreased. There has also been a decrease in the number of cases of unacceptably high ozone levels in Canada in the 1980s. There has, however, not been any significant decline in NOx levels in that decade. The air quality in Canada's major urban areas still leaves much to be desired. Even though more stringent automotive emissions standards have been introduced, there has been a great increase in the number of motor vehicles.

The situation is improving in Canada

In the latest annual air-quality report (April, 1990), the US Environmental Protection Agency gives some more good news. In the period 1979-1988: lead levels (see Chapter VII) decreased by 89%, SO_2 by 30%, carbon monoxide (see Chapter III) by 28%, particulates by 20%, and NO_2 by seven percent. However, smog levels increased by two percent, with an eight percent increase from 1987 to 1988, due in part to the hot summer of 1988.

Air quality is improving in USA

New regulations coming into force

The new initiatives for cleanup of air pollution that have been taken in North America and Western Europe suggest that these positive developments will continue. The new US Clean Air Act and the new regulations introduced in California are major steps towards improving the quality of outdoor air. It will be very interesting to see if people in the Los Angeles area can continue their love affair with the private automobile while still improving the quality of their air.

Air pollution is very serious in Eastern Europe

There is both bad and good news from Eastern Europe. Many people in the western industrial countries were shocked when in the late 1980s they recognized the severity of the air pollution problems in Poland, Czechoslovakia, East Germany and Romania. These problems are due mostly to pollution from heavy industry and high sulphur coal burning electrical stations. The good news is that something is now being done to reduce the gross air pollution from these sources. It will take many years and billions of dollars before the quality of air in parts of these countries improves to the level which we consider acceptable in Canada.

The situation in the Third World will get worse

The situation in Third World countries is not promising. Expanding industrialization and increased numbers of automobiles are causing serious air pollution problems in many major urban areas in these countries. Unfortunately, it seems unlikely that the situation will improve. These countries are nearly all burdened by heavy debts, and cannot readily undertake major air pollution clean-up campaigns. The governments often rely on more industrialization to provide employment for their rapidly increasing populations and to earn foreign exchange to pay their debts.

Pollution from burning oil wells in Kuwait

In early 1991 the world experienced one of the major air pollution catastrophes in all of history, the burning of oil fields in Kuwait. "Black rain" containing the smoke from these fires has been reported from a number of neighbouring countries. Most of the pollutants introduced by these fires will soon be washed out of the atmosphere, as they seem to have remained in the troposphere. However, they may cause an increase in carcinogenic PAHs (see Chapter III) in the region, resulting in higher cancer rates. The long-term effects of these fires remain to be assessed.

Recommended Reading

"Chemistry of the Atmosphere", R.P. Wayne. Clarendon Press, Oxford, 1991.

"Eastern Europe's Dark Dawn", J. Thompson. *National Geographic*, vol 79, No 6, pp 36-69, June 1991.

"Canadian Perspectives on Air Pollution". SOE Report No. 90-1, Environment Canada, 1990.

"Gaia: A New Look at Life on Earth", J.E. Lovelock. Oxford University Press, Oxford, 1979.

CHAPTER II
GLOBAL ATMOSPHERIC CHANGE

The atmosphere is not an infinite sink

Weather. It is a central concern in everyone's daily life; always changing, difficult to predict. We have long accepted that we cannot control the weather. But it seems that, without realizing it, our actions may be affecting the weather patterns of the earth. For centuries we have assumed that the atmosphere was a limitless repository for airborne chemical wastes. We now realize that our behaviour has been changing the composition of the atmosphere. Activities such as the burning of fossil fuels, massive deforestation and the use of aerosol sprays have been enlarging the layer of "greenhouse gases" that insulate the planet like a blanket. Because of this we may be causing changes in our global climate. The term "climate" is used to describe the pattern of average weather conditions over long periods of at least 30 years.

Upper levels of the atmosphere are affected

Some of these airborne chemicals eventually reach the earth's upper atmosphere where they pose a different problem. Certain chemicals, in particular the chlorofluorocarbons (CFCs), may seriously deplete the stratospheric ozone layer. This could affect life on earth since the ozone layer shields the planet from exposure to excessive amounts of ultraviolet (UV) radiation coming from the sun.

The severity of the effects are uncertain

The potential effects of global warming and ozone depletion are difficult to predict. As a result, there is a wide range of opinion of how seriously we will be affected by chemical changes in the atmosphere. There is still a possibility that the global warming trend and thinning areas of ozone observed by researchers are natural "blips" in natural cycles. It may be that the atmosphere can cope with the additional chemicals and the effects on our planet's surface will be very slight.

The effects can be extremely serious

Alternatively, global atmospheric change could be a serious problem. If the earth's climate is becoming warmer, the temperature increase will not be distributed equally across the globe. We could expect some regions to be severely affected while others experience little. Overall, we could expect more frequent and severe weather extremes and a shift in rainfall patterns. A reduced ozone layer could mean that all life on the planet would be vulnerable to the effects of increased exposure to UV radiation. These effects range from quite certain ones, such as an increase in skin cancers, to some potential subtle effects such as a depression in immune systems.

The effects are long-term

There is one certainty in the issue of global change: the atmosphere is slow to react. This means that there is a delay in the effects of past emissions. It also means that even if all fossil fuel and CFCs emissions were stopped today, the effects of any damage already done would continue for decades.

Solutions require lifestyle changes

Global atmospheric change is a very difficult issue. It is a problem which defies national or political borders; yet the consequences will be felt by each and every person. More difficult again, global change is caused by nothing short of our whole way of life. Reliance on fossil fuels underpins the economies of the industrialized nations; oil, coal and gas fire our cars and industry, generate our electricity, heat our homes. To seriously tackle the problem of global change means to accept that our energy-dependent and resource-wasteful lifestyles are causing most of the problems, and to alter some basic aspects of our lifestyles.

Global climatic change and ozone depletion are both complex issues, complete with contradictory opinions and controversial findings. We will examine them separately and in more detail in this chapter.

GLOBAL CLIMATIC CHANGE

The climate is always changing

The earth's climate results from complex interactions between the radiation from the sun, the atmosphere, land surfaces, the biosphere, and the oceans. It changes constantly, but the changes are so slow that climatic effects are not usually noticeable within our relatively short human lifespan.

Natural events affect climate

Some natural unexpected events can also occur which have more immediate effects on weather worldwide. Whenever dust is introduced into the atmosphere, for instance from volcanic eruptions, the amount of solar radiation which can reach the surface is reduced and the earth can be cooled as a result.

The earth reflects and absorbs sunlight

Later, when such dust falls to earth, the effect is reversed. Dark surfaces absorb sunlight better than light ones, hence, blackened snow and ice will enhance surface warming. The greater reflectivity of lighter coloured surfaces also means that deserts reflect more visible radiation than forests. Thus, when deforestation and desertification occur, the earth's surface becomes more reflective and the climate is cooled. Although deforestation and desertification occur naturally, human activities are probably responsible in most of the cases where these processes are occurring.

The Greenhouse Effect

Heating of greenhouses

While many factors affect global temperature, one of the most important in terms of global warming is the "greenhouse effect". This concept describes the principle underlying greenhouses, where the shorter wavelength visible radiation (see Appendix A) passes through the glass or plastic windows. It is to a large extent absorbed by the plants and soil, resulting in a heating effect. The plants and soil emit some of this heat radiation, but it has a wavelength in the infrared (IR). This does not readily pass through glass, but is absorbed and partially radiated back into the greenhouse. The temperature in the greenhouse thus remains warmer than the outside. This is also how the earth maintains its warm, hospitable climate in the extreme coldness of space.

The earth as a greenhouse

The earth is surrounded by a layer of gases, including carbon dioxide and water vapour, which insulate the planet much like the panes of glass in a greenhouse. Like these windows, the greenhouse gases allow shorter wavelength visible radiation from the sun to pass unaffected to the earth's surface. Some of this energy is absorbed by the earth, converted to heat energy and emitted as longer wavelength, infrared radiation (heat). The greenhouse gases, like the greenhouse windows, trap this reflected heat, preventing some of it from escaping to space. Without the "greenhouse effect" the earth would be at least 35 °C cooler, and life as we know it could not exist.

Appendix A describes electromagnetic radiation

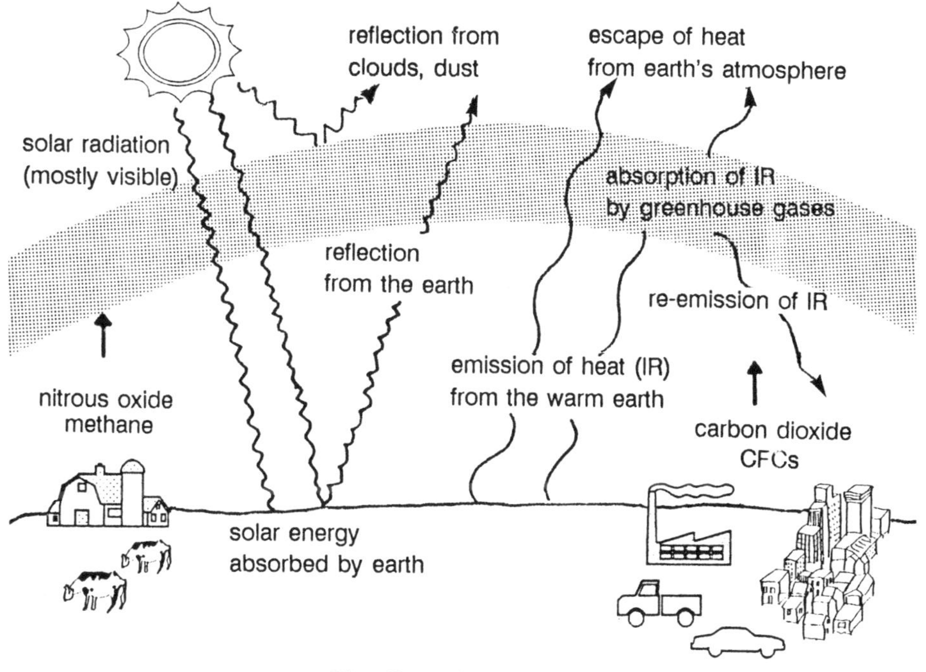

The Greenhouse Effect

Our activities are changing the atmosphere

We now recognize that human activities are increasing the concentrations of greenhouse gases in the atmosphere. Carbon dioxide (CO_2) is the main culprit, but we are also contributing considerable amounts of methane (CH_4), nitrous oxide (laughing gas, N_2O), low level ozone (O_3), and synthetic chemicals such as chlorofluorocarbons (CFCs) to the atmosphere. The increase in emission of these greenhouse gases are expected to amplify the earth's natural heat-trapping abilities, raising the temperatures in the lower atmosphere and on the surface. One major difference between natural climatic change and global warming is that with the latter we can expect significant temperature changes to occur within a century, rather than over thousands of years as in most natural change.

Impacts of Global Warming

Predictions can be made from historical data

It is difficult to predict the impacts of the effects of an increased burden of greenhouse gases on the world's climate. One way is to look at the past. Air from bubbles trapped in Arctic ice as long as 140 000 years ago can be tested for carbon dioxide levels. These measurements show that during the most recent ice age, levels of this gas were very low - 180 parts per million (ppm) as compared to the current 335 ppm. The correlation between low carbon dioxide levels and cold climate appears to support the theory that high carbon dioxide levels will warm the climate. But the use of such correlations is dangerous (see Chapter VIII). It may not be correct to assume that low carbon dioxide levels preceded and caused the ice age; the reverse may be true, or there may not even be a direct cause and effect relationship in this correlation. Even if the carbon dioxide level in the atmosphere was an important contributing factor in starting an ice age, it is unlikely to have been the only one.

Computer models of the global climate

Computer modelling is the predictive tool that researchers rely upon most heavily. Worldwide, there are at least 14 different groups studying climatic change using computers. Usually the studies examine the climatic consequences of doubling the atmospheric concentrations of greenhouse gases. All computer models predict an increase in average global temperatures, but the magnitude of predicted change varies considerably, ranging from 1.5 to 5 °C depending on the model used for the forecast. Researchers are trying to make these models more realistic by expanding them to include more factors which affect climate. For example, when the effect of clouds was incorporated into a computer model, the predicted average warming for the earth fell from 5.2 to 1.9 °C. Only now are scientists starting to include the effects of biological processes in their models.

Global warming can be large and rapid

Computer models predict climatic changes which are large and rapid relative to what has occurred in the past. An increase in the average global temperature of between 1.5 and 5 °C over a century may seem small, but its significance becomes clearer by noting that even during the last ice age, global temperatures were only about 5 °C lower than they are today. This is because a change in average global temperature is not distributed equally over the planet. The consensus among scientists studying global warming is that the temperatures would rise more around the poles than around the equator. Global warming would mean an increase in erratic, extreme weather events such as hurricanes and tornadoes. An overall increase in precipitation is also expected because a warmer planet will lead to increased evaporation. However, like the temperature increase, the precipitation would not be distributed evenly. Some areas would become wetter but others could become drier.

More impacts on agriculture

The nations that lie in the mid-latitudes are the world's richest food producing areas and global warming could have both beneficial and detrimental effects on this capability. It should be possible for agriculture to move northwards, following the climatic conditions best suited to specific foodstuffs. But on a human scale, such a change would entail severe social and economic dislocation. As well, the soils in some of these regions are not suitable for farming, so that intensive agriculture could not extend northwards in all cases. Disease and insect pests might spread northward following the warmer climates. Furthermore, longer growing seasons would provide longer breeding seasons for insects which might lead to larger pest populations and more massive infestations.

More storms in tropical regions

The tropical and sub-tropical regions of the world are expected to experience less warming. Instead, these areas would feel the effects of global change in the form of more extreme and more frequent tropical storms and droughts. Many developing nations lie within these climatic zones. These countries have long paid a high price, both in human lives and economic stability, as a result of natural weather disasters. We can expect this situation to worsen if the predictions for global warming are correct.

Rise in sea levels

Sea levels could be expected to rise 0.5 to 1.5 metres in the next century because the oceans would expand as they warm. Higher sea levels would cause an increase in the frequency and severity of floods, with serious consequences for many low-lying countries and coastal cities. One third of the world's population lives in low-lying coastal areas such as the Netherlands, the Maldives, the Pacific Islands, Bangladesh, Egypt, New York, Miami, New Orleans and Venice. The drinking water supplies of many coastal areas would also be threatened by salination as the world's river deltas became inundated with sea water.

More overall precipitation

An increase in the global temperature means an increase in the rate of evaporation of water from oceans, plants and soil. This means an increase in water vapour in the atmosphere, an increase in cloud cover, and an increase in precipitation. An increase in water vapour will increase the greenhouse effect, as water vapour is an important greenhouse gas. On the other hand, an increase in cloud cover will moderate global warming as sunlight is reflected from the tops of clouds. Recent evidence suggests that the reflection from clouds is a very significant effect. The predicted increase in precipitation would not be uniform. Some areas, such as California, may experience greater water shortages due to greater evaporation of their water supplies without a compensating increase in precipitation.

Implications for Canada

Many changes are ahead for Canada with global warming. Some changes would be beneficial, some neutral and others detrimental. We would need less fuel to heat our homes, but may increase the amount of air-conditioning. The Prairie grain belt could be extended northward, but this might be accompanied by a northward expansion of a dry belt into what are now major wheat producing areas. However, higher carbon dioxide concentrations may increase the yields of some crops such as alfalfa and wheat. As well, southern Ontario and Quebec would benefit from a longer growing season. The forestry industry could also benefit from a longer growing season, assuming the trees could adapt to the changing climatic conditions. Northern Canada would warm significantly, pushing areas of continuous permafrost further northward, and areas of discontinuous permafrost would melt gradually and irreversibly. This could have many repercussions on the delicate ecosystems of the North as well as on the lifestyles of the people who live there.

A new ice age may even be possible

There are suggestions that an increase in greenhouse gases will not lead to warming at higher latitudes. When more hot, moist air rises in tropical regions, cold polar air would move in to replace it. This would result in higher winds, and more violent hurricanes and tornadoes in tropical and sub-tropical regions, as in the common scenario for global warming. However, the consequences for polar regions could be quite different from the usual predictions. The tropical air currents could carry their moisture to higher latitudes where it would fall as rain, or snow. An increase in snowfall would cause the polar regions to experience increasingly long, cold winters; severe flooding in the spring; a slow buildup of glaciers; and, possibly, even another ice age.

THE GREENHOUSE GASES

Our activities add greenhouse gases

Many of the gases released to the atmosphere by our industrial society are greenhouse gases. But several of them warrant specific attention because of the quantities being released or the potency of these chemicals as greenhouse gases. Methane, nitrous oxide, low-level ozone and CFCs fit into one or both of these categories. But the carbon dioxide we emit to the atmosphere is usually considered to be the single largest contributor to an amplified greenhouse effect. Nearly half of the predicted global warming is believed to be due to carbon dioxide.

Carbon Dioxide

Carbon dioxide is essential for life

Carbon dioxide is essential for life on earth. Through photosynthesis and other metabolic processes, carbon dioxide is changed into different compounds which are the primary material from which all living organisms are made. In addition, although it accounts for only about 0.03% or 300 ppm of the atmosphere, carbon dioxide exerts a profound influence on the radiation balance of the earth. This gas has always been a major cause of the greenhouse effect, which has kept the earth's atmosphere at a habitable mean temperature of 15 °C.

Rates are in billions of tonnes of carbon per year

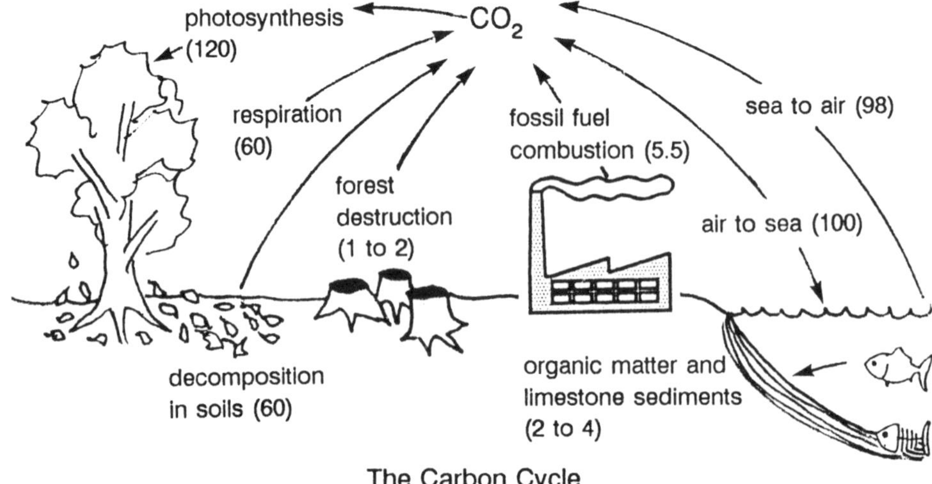

The Carbon Cycle

The carbon cycle controls CO_2 levels

The normal level of carbon dioxide in the atmosphere is determined primarily by the balance between what is given off by respiring plants and animals and what is taken up by photosynthetic organisms. This "carbon cycle", one of several "biogeochemical" cycles, is shown schematically above.

Fossil fuel use increases CO_2 emissions

However, our activities are disturbing the balance of carbon dioxide in the atmosphere. The combustion of fossil fuels (coal, oil and natural gas) employed in a multitude of activities including the generation of electricity, heating and transportation release large quantities of carbon dioxide to the atmosphere every day. In addition, the recent burning of massive tracts of tropical forests has liberated large amounts of carbon dioxide into the atmosphere. Once in the atmosphere, carbon dioxide has a lifespan of about seven years.

Atmospheric levels of CO_2 are rising

The increase in atmospheric carbon dioxide has been recognized for some time. This chemical has been monitored since 1957 when the first station was established on Mauna Loa, Hawaii. Between 1957 and 1980 the concentration of carbon dioxide went from 310 ppm to 335 ppm, an increase of eight percent. This increase has been confirmed by other stations, some of which are in Canada: Alert, Northwest Territories and Sable Island, Nova Scotia. Monitoring also showed that atmospheric concentrations of carbon dioxide undergo yearly cycles, peaking in late winter and reaching a minimum in the fall. This corresponds to the production of carbon dioxide by respiration and fuel burning in the winter and the surge of photosynthesis during the summer. Based on the data already gathered, the concentration of carbon dioxide in the atmosphere is projected by 2025 to be double what it was before the Industrial Revolution.

Carbon cycle can not absorb all extra CO_2

These increased quantities of carbon dioxide place a large burden on the biogeochemical cycle that processes carbon dioxide through the biosphere. It is estimated that 50% of the carbon dioxide produced by the combustion of fossil fuels remains in the atmosphere. The remaining 50% is presumably taken up by the oceans and photosynthetic plants both on land and in water.

Role of forests in the carbon cycle

A considerable amount of carbon dioxide is removed from the atmosphere by growing forests. At maturity, forests hold 10 to 20 times more carbon than the equivalent area of agricultural land. Deforestation by burning releases this carbon to the atmosphere. Humus and litter remaining from forestry operations also release carbon dioxide and methane as they decompose. The large tropical forests of the world are being encroached upon more and more, but this is compensated for to some degree by reforestation and regrowth of abandoned farmland in the Northern Hemisphere.

Role of oceans is not well understood

The role of the oceans in the carbon cycle is not well understood; oceanographers disagree on how much carbon dioxide the oceans can hold and how much is exchanged between the surface layers and depths. Carbon is incorporated into minerals which precipitate out in sediments, removing some of the carbon from the cycle for longer periods. Photosynthetic activity in the oceans is an important mechanism for removing carbon dioxide from the atmosphere, but it is also not fully understood.

Other Greenhouse Gases

Methane is a greenhouse gas

Almost half of the methane entering the atmosphere comes from the anaerobic (without air) bacterial decomposition of organic matter in the guts of sheep, cows, and termites, and in rice paddies where vegetation can rot in water-logged, oxygen-deficient soils. About 20% of global methane is produced from the incomplete combustion of wood or vegetation, which occurs in forest fires, land clearing and domestic fires. It is also released from the mining of coal and the production of natural gas, of which methane is the main component.

Methane is increasing rapidly

Methane is the greenhouse gas which is increasing most rapidly in concentration. There is about 1.7 ppm of methane in the atmosphere and it is increasing by one to two percent a year. Although there is less methane in the atmosphere than carbon dioxide, methane is 15 to 30 times more effective at reradiating infrared energy back to the earth. Methane has an average lifetime of about nine years in the atmosphere before it is oxidized to produce carbon dioxide and water vapour.

Nitrous oxide is another greenhouse gas

Nitrous oxide was discussed in Chapter I as one of the chemicals contributing to urban air pollution. But it is also a greenhouse gas and a cause of the depletion of the ozone layer. Nitrous oxide is formed naturally by microbes in soil and water. The major industrial routes for nitrous oxides entering the atmosphere are the breakdown of nitrogen-based chemical fertilizers and the burning of plant matter, including fossil fuels and deforestation.

N_2O is long-lived and increasing

Nitrous oxide has a lifetime of about 150 years in the lower atmosphere and it is about 150 times better than carbon dioxide at trapping heat. The only atmospheric decomposition process known for nitrous oxide is in the upper atmosphere where high energy ultraviolet radiation (see Appendix A) alters nitrous oxide, making it possible for it to form nitric oxide. The concentration of this gas has increased to 0.30 ppm from 0.28 ppm in the early 1900s.

Ozone is a greenhouse gas

Surface ozone (or tropospheric ozone) was introduced in Chapter I as a constituent of smog formed by photochemical reactions between sunlight and vehicle emissions such as carbon monoxide and nitrogen oxides. As a greenhouse gas, surface ozone is short-lived with a lifespan of about two to three weeks, but it is roughly 2,000 times more effective as a greenhouse gas than carbon dioxide. Tropospheric ozone levels over North America and Europe may have increased one to two percent annually in the past decade.

CFCs are part of the greenhouse effect

The amounts of CFCs in the lower atmosphere are increasing at an alarming rate. CFCs are about 10,000 times more powerful as greenhouse gases than carbon dioxide. They are used as aerosol propellants, blowing agents for foam insulation, coolants in refrigerators and air conditioners, and solvents for cleaning electronic microcircuits. CFCs contribute to both the greenhouse effect and the depletion of the natural ozone layer in the upper atmosphere. Restricting the usage of CFCs would thus help address two problems. We examine CFCs in detail during the discussion of the ozone hole later in the chapter.

ACTION ON GLOBAL WARMING

Options for coping with global warming

Scientists and government officials have suggested several options to deal with the conflicting ideas on the consequences of the greenhouse effect. Some feel that the future warming of the earth is inevitable unless immediate cutbacks of greenhouse gases are initiated to prevent or at least curtail the warming. Others believe that since many greenhouse gases are emitted through common human activities such as growing rice and driving cars, it will be too difficult to limit them, and global climate change will be inevitable. Thus time and energy should be focused on thinking up ways in which humans can adapt to the warming. Finally, there are those who think that no decision can be made until more conclusive facts and evidence are produced.

International panel on climate change

An "Intergovernmental Panel on Climate Change" (IPCC) was set up in 1988 to prepare reports on various aspects of the global warming issue. Britain heads a section devoted to scientific research, the USSR is committed to forecasting the effects on global climate and the US leads a panel on response strategies to global warming. In June 1990, the complete scientific research report formulated by over 300 scientists from over 40 countries was the basis of an international conference in London. The report concludes that without reductions in emissions of greenhouse gases, global mean temperature will rise by 3 °C by the year 2100.

Responses to the IPCC report

In response to the IPCC's report, the British government has committed to reducing Britain's currently projected carbon dioxide emission levels by some 30% by the year 2005. Many IPCC scientists think that this target will merely "buy time". The US, Japanese and Canadian governments are reluctant to reduce carbon dioxide emissions in these nations because of the economic ramifications and what they consider to be a lack of sufficiently convincing evidence of the harmful effects of global warming.

Ideas for reducing CO_2 emissions

A number of policies have been proposed for limiting the increase in concentrations of greenhouse gases in the atmosphere. Carbon dioxide emissions could be reduced through general energy conservation measures which would mean less burning of fossil fuels. There could be more development of energy sources such as solar, wind, tidal, geothermal and nuclear. These sources could replace fossil fuel sources, or at least slow the increase in the use of fossil fuels. There have been suggestions of a "carbon tax" to discourage the use of fossil fuels.

Removal of CO_2 by trees

Reforestation would increase the removal of carbon dioxide from the atmosphere. It should be noted that a mature forest stores a great deal of carbon, but does not result in any net removal of carbon from the atmosphere. There have been cases where on the start-up of a new coal-fired electrical station (eg in the Netherlands), a commitment has been made to reforest an appropriate area of tropical land so that there would not be a net increase of carbon dioxide into the atmosphere. Research programs in agriculture may find means of reducing methane emissions.

Technological fixes are unwise

Some exotic "technological fixes" have been proposed as solutions to global warming. One such suggestion is to dump iron into the oceans to supply nutrients to marine plants, thus increasing photosynthetic activity in the oceans. Such massive technological interventions seem unwise, as they may lead to even greater problems that we cannot foresee with our limited knowledge.

STRATOSPHERIC OZONE DEPLETION

Stratospheric ozone is decreasing

Since the 1970s, scientific attention has focused on the depletion of the "ozone layer". Scientists do not fully understand the changes which are occurring in the upper atmosphere, but they are concerned about the consequences. Ozone protects us from damaging ultraviolet rays of the sun. Exposure to this type of radiation (see Appendix A) can cause skin cancer, reduce crop yields and damage aquatic life. Some changes in the ozone layer could be naturally caused, but evidence is now pointing to industrially produced chemicals, namely chlorofluorocarbons (CFCs), as the major cause of ozone depletion. For this reason, the discussion of ozone "holes" is followed by a review of CFCs and their place in today's society.

The Stratosphere

The atmosphere consists of layers

The temperature, pressure and composition of the earth's atmosphere is not constant, but varies with altitude (see below). The lowest layer of the atmosphere, the troposphere, extends from the earth up to about 11 km. It contains almost all atmospheric water vapour and comprises 80% of the atmospheric mass. This is a region of swirling air masses, warm and cold fronts and cloud formations; it is the zone in which weather occurs. It is marked by a temperature decrease of about 6.5°C per kilometre altitude due to the increasing distance from the sun-warmed earth. The tropopause at the top of the troposphere is a narrow region of nearly constant temperature, about -50 °C. The stratosphere extends about 50 km above the tropopause. The atmosphere warms from -50 to about 0°C in this region due to the presence of ozone.

The regions of the earth's atmosphere

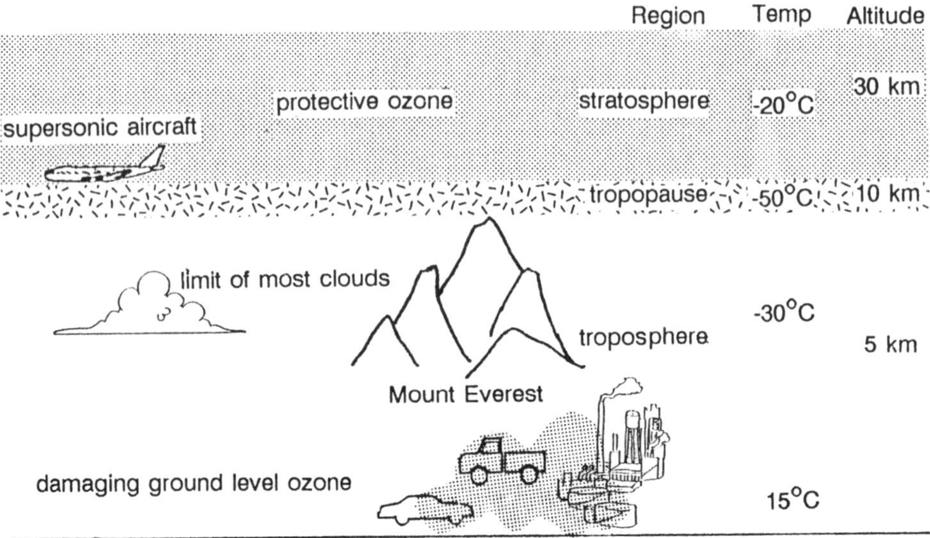

The ozone layer referred to in this pollution issue is distinct from the surface ozone discussed in Chapter I as an element of smog. The ozone layer is a naturally occurring band of ozone about 20 km thick, located in the stratosphere about 15 to 35 km above the earth. However, since the pressure of the atmosphere at such altitudes is very low, the ozone is distributed so thinly that it would collapse to a three mm thick wide band of pure ozone if it were at normal atmospheric pressure. It is this stratospheric ozone which is very important to all life on earth.

Most ozone is in the stratosphere

The ozone layer acts as a protective shield, absorbing harmful ultraviolet UV-B radiation (see Appendix A). There is a strong connection between UV-B radiation and incidence of non-melanoma skin cancer in humans. Scientists believe that there is also a link between this type of radiation and melanoma skin cancer, which is more likely to be fatal. It is argued that a one percent reduction in stratospheric ozone would result in a two percent increase in both forms of skin cancer. To put these numbers into perspective, this increase in skin cancer is equivalent to the US population receiving the extra UV-B obtained by moving south by about 110 kilometres.

Ozone protects us from UV-B

The natural ozone concentration in the stratosphere is kept fairly constant since the rates of formation and destruction of ozone are roughly equivalent. When extra ozone destroyers are introduced into the stratosphere, the amount of ozone drops to a lower steady state value. The new rates of formation and destruction are again balanced and the amount of ozone is steady, but it is now less than it was originally.

Ozone is naturally in a steady state

This reduction of the steady state concentration of ozone is what sparked the outcry against the development in the US of Supersonic Transports in the 1970s. It was discovered that these aircraft, which would be flying in the stratosphere, emitted exhaust fumes containing ozone-destroying nitrogen oxides. The subsequent development of this kind of jet in Europe (the Concorde) proved uneconomical, and only a few Concordes are in use. A post-Concorde aircraft currently being developed by Britain and France poses the same environmental problems.

Concerns about supersonic transport

The Effect of Chlorofluorocarbons

In the mid 1970s attention was focused away from these aircraft to the household spray can. Industrial chemicals called chlorofluorocarbons (CFCs) were used as the propellant in many such cans. Scientists became concerned that these chemicals could harm the ozone layer when they realized that CFCs were building up in the atmosphere. CFCs are very stable in the lower atmosphere, so that they persist for a long time. Eventually they diffuse into the stratosphere, where the CFC molecules can absorb high-energy ultraviolet radiation, which causes the molecules to split into chlorine atoms and other products. It is the chlorine atom which is the source of the problem. Through its action as a catalyst, a single chlorine atom is capable of destroying thousands of molecules of ozone.

Concerns about spray cans

The effects of CFCs are delayed

An important aspect of the ozone depletion problem is the time delay between the release of CFCs into the lower atmosphere and their arrival in the stratosphere. Since decades are required for CFCs to diffuse from their release sites to the stratosphere, any decision taken today to ban CFCs will not be reflected in the stratosphere for some time to come.

Other industrial chemicals affect ozone

CFCs are not the only industrial chemicals which could harm the ozone layer. A closely related class of chemicals called halons also destroy ozone. Halons are substances containing bromine which are commonly used in industrial fire extinguishers. They are not as common as CFCs, but they are ten times more effective than CFCs at depleting ozone, as bromine is a more effective catalyst than chlorine in breaking down ozone. Widely used organic solvents such as methyl chloroform (1,1,1-trichloroethane, CH_3CCl_3) and carbon tetrachloride (CCl_4) also are believed to significantly contribute to stratospheric chlorine levels.

Natural chemicals also affect ozone

Natural chemicals can also affect stratospheric ozone. Nitrogen fertilizers have been implicated in ozone destruction. The claim has been made that nitrous oxide production is accelerated by these fertilizers to such an extent that the ozone layer could be depleted by as much as 15% over a century. Oceans, volcanic eruptions, and burning of biomass also release chlorine compounds into the air which could contribute to the destruction of ozone. The increased production of methane and carbon dioxide may, however, increase the production of ozone in the stratosphere. There is currently no conclusive answer to the question of how seriously CFCs affect stratospheric ozone.

It is difficult to measure ozone depletion

A major problem in dealing with the issue of ozone depletion is the measurement of that depletion. The concentration of ozone in any one place varies daily and seasonally and is believed to follow the 11-year sunspot cycle. Volcanic eruptions sending material into the stratosphere may also affect ozone levels there. Any change in ozone levels caused by industrial activity would have to be of the order of ten percent before it could be noticed against the natural variation.

The antarctic "ozone hole"

In 1985, an "ozone hole" was discovered in the atmosphere above Antarctica. The word "hole" is misleading, since the region is actually an area of lower concentration of stratospheric ozone, not a complete absence. The hole is also not a constant entity. It occurs annually in the polar spring during the months of September and October; then it drifts away from Antarctica and eventually breaks up, mixing with air at lower altitudes. In October 1987, the ozone hole had an area of the size of the continental United States and persisted until early December. It was considerably smaller in 1988, but in 1989 it again increased.

The causes of the holes are not well known

The exact cause of these "holes" is unclear. Some Soviet scientists argue that they may be natural, but many scientists believe that CFCs and other ozone depleting substances are implicated. Scientific expeditions in 1987 and in 1989 found the polar atmosphere to have high levels of chlorine compounds. This discovery links CFCs and other chlorine releasers to ozone depletion.

Health and Environmental Effects

The immune system may be suppressed

If the stratospheric ozone layer is being destroyed, it could have serious repercussions on living things on the planet. A United Nations report suggests that ozone depletion may increase the number and severity of outbreaks of infectious diseases such as measles, herpes, tuberculosis and leprosy. Normal immune systems seem to become less effective when exposed to large doses of UV-B radiation. Increases in UV-B are expected to reduce the effectiveness of vaccinations, and increase the incidence of eye cataracts. For every one percent depletion in ozone, an extra 100,000 people would become blind.

Air pollution may increase

Stratospheric ozone depletion may result in an increased pollution of air near ground level. Larger amounts of UV-B radiation reaching the troposphere will create increased quantities of reactive chemicals, such as hydroxyl radicals (OH); these are reactive because they contain a lone electron which is ready to form a chemical bond. Such chemicals could cause the formation of tropospheric ozone and other damaging pollutants like hydrogen peroxide and acids. Hydroxyl radicals are, however, effective in reducing methane, a harmful greenhouse gas. UV-B radiation also increases the rate of degradation of plastics, rubber, wood, textiles and paints.

Ozone depletion could affect climate

A decrease in stratospheric ozone could also cause alterations in the structure of the atmosphere, eg the tropopause could move to a lower altitude. This could have considerable effects on global weather patterns.

Can we eliminate CFCs?

There is thus some evidence, although not completely conclusive, that CFCs may cause detrimental changes to the planet's atmosphere. Can we do without these industrial compounds and eliminate their use? This is discussed in the next section.

CHLOROFLUOROCARBONS

Development of CFCs as refrigerants

In the 1920s, the refrigeration industry was trying to find replacements for the toxic and corrosive compounds sulphur dioxide and ammonia then in use as refrigerants. Almost by accident, it was found that the compound dichlorodifluoromethane (CCl_2F_2, a chlorofluorocarbon, or CFC) was non-corrosive, had the correct thermal properties, and had a very low toxicity. By 1929, an experimental plant had been built to produce these chemicals.

CFCs have a variety of uses

A large variety of CFCs were developed for use in air-conditioning, refrigeration, equipment sterilization, fire suppression, and in medical inhalation products. CFCs were considered a very desirable development because they were very stable, non-flammable, and non-toxic. Today in Canada it would be very hard to go through a day without using products containing CFCs. They are used as blowing agents in the manufacture of polyurethane foams, which are used in such things as automotive seating, mattresses, carpet padding, thermal insulation and refrigerated trucks. CFCs are also used to clean airplanes, computer chlps, and communication equipment. Some aerosols such as hairspray and underarm deodorant used to contain CFCs as a propellant but this use was banned in Canada in 1979.

Environmental fate of CFCs was ignored

During the decades of increasing CFC use, no serious thought was given to the eventual fate of the substances emitted from aerosols, fire extinguishers, or scrapped refrigeration and air-conditioning systems. In 1971, measurable traces of CFCs were found in the atmosphere. The CFCs are unreactive and essentially insoluble in water. Thus there does not seem to be any natural "sink", such as a chemical degradation or cleansing system, for these chemicals in the lower atmosphere.

CFCs could reach the stratosphere

In 1974, two scientists published the conclusions to their speculations about the eventual fate of the CFCs being released into the atmosphere. They suggested that these substances would eventually diffuse into the upper atmosphere. This process would be slow, so that CFCs would reach the stratosphere, about 20 kilometres up, decades after release at sea level. This gas is chemically inert in the troposphere, but in the stratosphere it is decomposed by the more energetic solar radiation present there, and the fragments may catalyze the destruction of stratospheric ozone.

Production of CFCs fluctuates

Between 1960 and 1974, world production of CFCs increased by a factor of eight. Following the discovery of CFCs in the atmosphere in the early 1970s, several countries started to regulate CFCs. For example, USA, Canada, and the Scandinavian countries banned the use of CFCs in most aerosols by 1978 or 1979. Production of CFCs then actually fell during the years 1975 to 1982. This decline was partly due to regulations limiting the use of CFCs in aerosols, partly due to the public's growing environmental awareness, but also probably partly due to an economic recession, and an attitude that aerosols are non-essential or luxury items. Whatever the prevailing cause, the use of aerosols had thus halved by 1983. However, from 1982 to 1987, world production of CFCs rose again, mainly because of increased use in refrigeration and air-conditioning equipment.

Restrictions on CFCs

The "Montreal Protocol"

An agreement limiting the production of five CFCs and halons used as fire suppressants was reached in a United Nations Environment Programme (UNEP) conference in Montreal on 16 September 1987. The "Montreal Protocol" has since been ratified and has come into force. This agreement must be considered an optimistic step forward for the world; UNEP's Executive Director stated: "UNEP considers this accord unique because it seeks to anticipate and manage a world problem before it becomes an irreversible crisis".

Wealthy nations will help poor ones

The Montreal Protocol is also unique in that it is the first such agreement whereby rich countries have agreed to help out poor countries. A fund is to be set up to help developing countries buy ozone-safe technology. Some developing countries like China and India had threatened not to sign the accord without this help. Large developing countries could easily emit enough CFCs to negate any decreases in emissions from industrialized countries. The mechanics of the compensation will no doubt be discussed in 1992, when the next review of the protocol is scheduled.

Additional agreement on CFCs in 1990

In June 1990, an international agreement was signed which supersedes the Montreal Protocol. Under this accord, 81 countries will stop producing CFCs by the year 2000, with cuts of 50% by 1995 and 85% by 1997. Halons will also be phased out by the year 2000, with a 50% cut by 1995. Medical and other life saving uses of halons and CFCs may still be allowed as long as there are no alternative substances. Two other ozone depleting chemicals were added to the protocol: carbon tetrachloride and methyl chloroform. Carbon tetrachloride use will be reduced 85% by 1995 and completely by 2000. Methyl chloroform production will be cut to 70% of current levels by 2000 and phased out completely by the year 2005. A non-binding resolution was also approved for the phaseout of certain CFC substitutes, called hydrochlorofluorocarbons (HCFCs) by 2020 or 2040 by the latest.

Alternatives for CFCs

It is relatively easy to find alternatives for CFCs for aerosols, or to discontinue use of aerosols altogether. It is more difficult to find a replacement for CFCs in refrigeration. Refrigerators and air-conditioners may have been luxury items in previous decades, but they are now considered as essentials by many consumers. Refrigeration is vital for human health in much of the world due to its use in food preservation. It is difficult to find suitable alternatives for CFCs for these uses, but helium and propane are being considered. Two other types of chemicals, HCFCs and HFCs were created to replace CFCs. HFCs (hydrofluorocarbons) do not contain chlorine; thus they do not break down ozone. HCFCs (hydrochlorofluorocarbons) contain hydrogen in addition to the chlorine and are thus expected to break down more readily in the lower atmosphere. HCFCs are used in some refrigeration systems, aerosols, foamed polystyrene food packaging and in other foamed plastics.

Problems with replacements

There are, however, problems with HFCs and HCFCs. They are greenhouse gases, and according to the United States Environmental Protection Agency, one type of HCFC already contributes about four percent of the chlorine concentration in the atmosphere.

Methyl chloroform, the solvent used for cleaning metals in such things as airplanes and electronic components, may also not be easy to replace. Alternatives for some of its uses are already known, such as using soapy water to clean airplanes. However, many of the current alternatives are flammable, toxic or much more expensive.

It is a long-term issue

Phasing out CFCs will be expensive. About 135 billion dollars worth of machinery and equipment in the United States alone relies on CFCs. A recent study for America's Department of Energy predicted that a phase-out of CFCs by the year 2000 could cost anywhere from 19 billion to 34 billion dollars.

CONCLUDING REMARKS

Both issues have global implications

The two major issues discussed in this chapter have several points of similarity. They are both global in nature. The release of greenhouse gases or ozone depleting substances in any country will affect the whole world. Thus any agreement to control these chemicals must be international in scope.

They are long-term issues

The reasons why these problems are global is that they are long-term and affect our common atmosphere. The emissions from Canada remain sufficiently long in the earth's atmosphere to diffuse throughout it and be a problem common to all the planet's inhabitants. This long-term nature means that harmful effects are delayed, making it difficult to definitively prove that they exist. We must also wait for several years before any improvements will be seen if we succeed in controlling emissions.

Uncertainty about consequences

The delay between emissions and effects makes it difficult to be certain that the negative effects actually are occurring. Hypotheses, theories and models provide most of the evidence that there will be detrimental change in the global climate or that UV-B will increase. There is little direct evidence based on observations. It is more difficult to agree on measures to control chemical emissions if there is little direct evidence of harm. There are significant numbers of people who doubt the validity of the theories and models predicting serious harm to human health and the environment.

How safe is "safe enough"?

There may be some doubt about the evidence for harm from global atmospheric change, but there is a good possibility that such harm can be very substantial. The decisions to be taken in these cases illustrate the "planner's dilemma". How safe is safe enough? How much "insurance" should we buy to protect against further losses which are not completely certain but potentially great?

Do we value the global environment?

Our decisions depend to a great extent on how much we value the future, and how much we value the global environment as compared to our own present well-being (see Chapter VIII for a further discussion). Will we be prepared to make some personal sacrifices today, for example: conserve energy, pay more for a refrigerator? If we place a high value on the future world environment, then we should be ready to alter our personal lifestyles and national priorities appropriately.

Dilemma for the Third World

The industrialized countries may have quite a different attitude towards this than countries trying to industrialize (the Third World). We have to a large extent satiated our desires for material goods and convenience, using up a disproportionate amount of global resources in the process. It may be easier for us to stabilize or reduce emissions of carbon dioxide and even eliminate the use of CFCs. Many people in the Third World aspire to the same level of material comfort that we already have achieved. They will not be as willing to limit their use of resources for the sake of slowing global atmospheric change. When considering "sustainable development", the industrialized countries are likely to emphasize "sustainable"; in the Third World, many people still place "development", in particular economic development, as their top priority.

There are, however, some differences between the two main issues in the chapter. The problem of stratospheric ozone depletion is easier to deal with because the phasing out of CFCs is relatively easy. Uses can be discontinued or substitutes found without excessive disruption of industry or economic consequences. This is why an international agreement on CFCs has been possible.

Ozone depletion is easier to solve

In the case of greenhouse gases, especially carbon dioxide, any reduction means serious changes in industrial and personal activity. It involves decisions such as reduction in electricity generation, in commercial and personal transport or changes in farming activity. These are difficult decisions to make, especially if the consequences of the greenhouse effect are uncertain and far in the future.

Solutions to global warming

An international "Law of the Atmosphere" governing global atmospheric changes is very desirable. It may take many years before it is agreed upon, but the success of the Montreal Protocol shows that such agreements are possible.

Hope for an international agreement

Recommended Reading

"How Plankton Change the Climate", P. Williamson and J. Gribbin. *New Scientist*, pp 48-52, 16 March, 1991.

"Planet Under Stress: The Challenge of Global Change", C. Mungall and D.J. McLaren, eds. Oxford University Press, Don Mills Ont., 1990.

"Scientific Assessment of Climate Change", Report by Working Group 1 of the Intergovernmental Panel on Climate Change. World Meteorological Organization and United Nations Environment Programme, June 1990.

"Special Issue: CFCs and Stratospheric Ozone". *Ambio*, October 1990.

"Special Issue: Managing Planet Earth", Scientific American, September 1989.

"Our Common Future", World Commission on Environment and Development, Oxford University Press, Oxford, 1987.

CHAPTER III
INDOOR AIR POLLUTION

People breathe indoor air most often

Examine an average "day in the life" of a person in the developed world: she awakes in a house, travels from building to building as her work or pastimes dictate, and ends up back in her home at night, ready to start the cycle anew. People in northern, industrialized nations spend upwards of 90% of their time inside buildings. While we struggle to clean up the air outside, the air that we breathe indoors might be a more immediate threat to our health.

Office buildings can be polluted

The problems of indoor air pollution are very apparent in office buildings, where workers spend their days breathing recirculated air, reliant upon the ventilation system for periodic infusions of fresh air. The health complaints being voiced by office workers are numerous enough that the phenomenon has earned the name "sick building syndrome".

Airtight houses can trap pollutants

In non-industrial settings, poor indoor air quality seems to stem largely from efforts to conserve energy. Many of these efforts were stimulated in the 1970s by the rapid increase in the cost of energy, especially oil. Making modern buildings increasingly airtight with controlled ventilation has gone far toward conserving energy. It has also created a whole new air pollution problem, since airtight buildings trap not only heat but also airborne chemicals and dust.

Many sources of indoor air pollution

Indoor air pollutants arise from many sources. Cooking and bathing tend to create moist, humid conditions well suited to the growth of moulds and bacteria. These can stimulate reactions like "humidifier fever", an allergic lung disease caused by inhaling the microbes that can exist in humidifier water. Gas-burning stoves or heating systems produce substances like nitrogen oxides, carbon monoxide and other combustion by-products. In a study of 8,000 children in the US, evidence suggested an increased incidence of respiratory disease in homes using gas burning stoves.

Second-hand tobacco smoke

Cigarette smoking also pollutes the air. Smoking is already ranked as the number one preventable cause of death in North America. Since 1986, it has also been recognized to have serious health effects on non-smokers.

Asbestos fibres and radon gas cause concern

In the early 1970s, increasing awareness of the dangers of airborne asbestos fibres prompted widespread public concern. Extensive control and corrective measures were initiated to reduce or eliminate those risks. More recently, there has been concern over the problem of radon gas. Radon is an invisible, odourless gas that can seep through soil and masonry into buildings. It is natural, radioactive, carcinogenic, and prevalent in many areas of Canada.

SICK BUILDING SYNDROME

Complaints of poor health at workplace

In 1979, employees at the Canadian government office building, "Les Terrasses de la Chaudiere", came to the conclusion that the air in their building was literally making them sick. This grand structure accommodating more than 6,000 workers was intended as a showcase when it opened in 1978. Instead, many of the people who worked there began to think of it as a showcase for "sick building syndrome". People were complaining of symptoms such as undue fatigue, sore throats, nausea, chronic sneezing and colds, all of which seemed to diminish with time away from the building. The employees organized and began to pressure the government to take action.

Reasons why buildings are sick not known

The symptoms noticed by the employees at Les Terrasses are all commonly associated with the sick building syndrome. Unfortunately, employers have been reluctant to deal seriously with such complaints. Sick building syndrome is a mysterious issue. Although the sources or symptoms can vary, the problem is real. Complaints of poor health in "sick" buildings are becoming more widespread and there is a growing demand for comprehensive regulation of indoor air quality.

It is not a psychological problem

When this syndrome first surfaced in the mid 1970s, many researchers, for lack of a better explanation, assumed they were dealing with a psychological problem. As the number of complaints mounted, it became evident that the problem was not solely psychological, and now most investigators reject such an explanation for the sick building syndrome.

Microbes may contaminate indoor air

Biological contamination is one possibility being considered to account for the ill-effects on people in sick buildings. Bacteria, moulds and other microbes can induce allergic reactions in sensitive people or actively infect people, lowering their immunity to other illnesses like colds or flu. The notorious "Legionnaire's disease" that killed 29 hotel guests in Philadelphia in 1976 has reappeared across the North American continent and is generally believed to result from fungal contamination of air-conditioning systems.

Outdoor chemicals are unlikely sources

Chemicals from outdoor air have also been considered as contributors to sick building syndrome, but there is little evidence that they are a major cause. Office buildings, hospitals and schools seem to contain higher levels and a wider variety of toxic organic chemicals than outdoor air, even in heavily industrialized areas. People who live in areas heavily involved in petrochemical, plastics or paint processing have no greater incidence of sick building syndrome than those who live in less industrialized or even rural areas. This implies that the primary source of pollution is within the buildings themselves.

Office air can have many VOCs present

Building materials, cleaning solvents and furnishings all contain chemicals which can evaporate into the air; these are collectively called volatile organic compounds (VOCs). Some of the highest emitters of these chemicals are latex caulking, telephone cable, carpet adhesive and particle board. In total, more than 500 different VOCs have been found in tests of office air. The most commonly found compounds are aliphatic hydrocarbons (gasoline contains many aliphatics), followed by aromatic hydrocarbons like benzene, and then chlorinated hydrocarbons. Unfortunately, how each contaminant might affect people is not completely known. An interesting point that must be explained is that the symptoms being observed by people are immediate and tend to subside once they have left the "sick" building. The effect of long-term chronic exposures to these contaminants is still largely unknown.

New buildings have higher VOC levels

The levels of VOCs appear to be highest in new buildings, dropping noticeably within the first year after a structure's completion. Scandinavian countries require a greater exchange of air in a building with outdoor air for the first year of a building's life. In Canada, a federal/provincial committee is working on exposure guidelines for residential indoor air quality, and research is now in progress on the problem of office air specifically. In the United States, no single agency is responsible for regulation of indoor air quality, although the Environmental Protection Agency (EPA) is charged with studying the problem. While the EPA is shying away from instituting regulations because they do not feel they know enough about indoor air pollution, the states of Washington and California are going ahead with legislation to attempt to deal with the problem.

Scant knowledge prevents strong laws

Until more is learned about the causes and effects of indoor air pollution, preventing the formation of the pollutants is not very feasible. In the interim, technologies like air scrubbers and pollutant sponges have been proposed as ways of controlling indoor air pollution. But, as remedies go, the most effective one to date is simply more exchange with fresh outside air.

Urea-Formaldehyde Foam Insulation (UFFI)

UFFI caused problems in Canada

A specific example of sick-building syndrome is that of urea-formaldehyde foam insulation (UFFI). In the 1970s, UFFI was placed on the list of insulation materials which homeowners could install in the walls of their houses with a Canadian Government subsidy. In 1980, by which time it had been installed in over 80,000 homes, Health and Welfare Canada placed a temporary ban on UFFI. A number of people had complained of health problems after having UFFI installed in their homes.

Formaldehyde may affect health

UFFI is installed by pumping a resin and an acidic hardening agent together with compressed air into the cavities of a wall. When it is pumped in, the material looks and feels like shaving cream, but it soon hardens through a chemical reaction to a solid foam. The foam can become a site of fungal growth; spores of the fungus can thus infiltrate the house. The foam can also give off some formaldehyde, especially if it is hot and moist. At the time of the ban, formaldehyde was believed responsible for the health effects in UFFI homes.

Formaldehyde is present in normal air

Formaldehyde is present naturally in outdoor air, with concentrations ranging from 0.005 to 0.06 parts per million (ppm) depending on location. Formaldehyde is also present in our bodies. It is a normal metabolite of cells, and is active in the formation of amino acids and other chemicals essential to life.

Many sources and uses of formaldehyde

There are many sources of formaldehyde other than UFFI. It is present in smoke from all kinds of combustion: cigarettes, fire places, car engines, forest fires. It is also given off by a variety of products such as textiles, glues, plywood and electrical parts. The smell of formaldehyde is well known in biology laboratories and funeral homes from its use as a preservative and embalming fluid.

Affected homes get relief in Canada

In general, no ill-effects have been reported in homes with formaldehyde levels in air below 0.05 ppm. In a 1981 survey, about 25% of the houses containing UFFI were found to have a level above this; 5.1% had a level above 0.1 ppm. Some houses that contained no UFFI also had high formaldehyde levels; 2.6% had a level above 0.1 ppm. The Canadian government decided to provide up to $5,000 for remedial action to homeowners who had installed UFFI and whose homes had a level greater than 0.1 ppm of formaldehyde, or who could show adverse health effects due to formaldehyde.

Health effects of UFFI are still unclear

There is no definitive evidence that formaldehyde has been responsible for sick-building syndrome in "sick" UFFI houses. No excess cancer deaths were noted in major epidemiological studies (see Chapter VIII) of people occupationally exposed to high levels of formaldehyde. However, animal tests have linked formaldehyde to nasal cancer in rats and mice. There still seems to be no clear answer as to whether formaldehyde exposure has been the cause of any illness or death in houses with UFFI or in occupational settings.

SECOND HAND SMOKE

Smoking is a leading cause of death

Over the past 30 years, people have become increasingly aware of the dangers posed by smoking cigarettes. More than 50,000 scientific studies worldwide have firmly established tobacco smoke as a cancer-inducing substance. At least 60 carcinogens have been identified in tobacco and tobacco smoke. Cigarette smoking is believed to be the leading preventable cause of death in the US.

Smokers are buying fewer cigarettes

In Canada, cigarette smoking contributes to the deaths of roughly 30,000 people each year. Despite widely publicized health hazards and mounting societal pressure to stop, Canadians continue to smoke an average of about 2,500 cigarettes per person per year. The good news is that cigarette sales in Canada decreased by five percent in 1990, after a drop of 6.8% in 1989. Since 1984, the per capita use of tobacco in Canada has fallen by 28%.

Second-hand tobacco smoke raises concern

It can be argued that smoking is a personal decision, and that as long as people are made aware of the dangers, then it is their right to choose to smoke as long as they do not affect others. However, the health risks to non-smokers from exposure to "second-hand" tobacco smoke in the home and workplace are much more serious than previously realized.

Dangers to non-smokers is shifting attitude

The 1986 Surgeon General's Report in the US was the first in identifying a risk of disease from exposure to tobacco smoke for individuals other than smokers themselves. A recent US study concludes that second-hand smoke may cause 53,000 deaths each year (comprised of 3,700 lung cancers, 12,000 other cancers, and 37,000 heart disease victims). If these numbers are correct, involuntary smoking would become the third leading preventable cause of death in North America, after alcohol and smoking itself. Facts like these have added more fuel to the debate over smoking versus public health and may hasten a shift in public attitudes to view smoking as a privilege rather than a right.

Chemical Composition

Carbon monoxide is found in smoke

Second-hand smoke is a combination of the sidestream smoke given off at the tip of a burning cigarette and the smoke exhaled by a smoker. Both sidestream smoke and mainstream smoke consist of a combination of gases and very small particles containing over 3,800 different chemicals. About 90% of tobacco smoke is made up of gases. The major toxic gas is carbon monoxide. Other toxic gases emitted in tobacco smoke include formaldehyde, ammonia, benzene, nitrogen oxides and some potent carcinogens such as nitrosamines. Most of the known or suspected cancer causing agents, however, are found in the particles given off by burning tobacco. About 5 billion particles are present in a thimble-full of smoke. It has been estimated that 30 - 40% of these particles become embedded in the alveolar regions of the lung and 5 - 10% in the bronchial regions. These particles contain many known carcinogens such as the polyaromatic hydrocarbons (PAHs). They may take days or months to clear from the lungs.

Tobacco tar is a carcinogen

The tar in tobacco smoke is a "complete carcinogen". It can even cause cancer when it is put on the skin of animals. It seems to contain tumour promoters, initiators and accelerators - all the ingredients involved in the development of cancer. Tumour initiation is due to the (PAHs), such as benzopyrene. Such PAHs are present in tobacco smoke, but also in the smoke from chimneys, automobile exhausts and other combustion processes where the combustion may be incomplete. PAHs have been identified in soils near highways.

Risk to non-smokers is not yet clear

There is some difference in the chemical constituents of mainstream and sidestream smoke. Because sidestream smoke is formed at a lower temperature, when the smoker is not puffing on the cigarette, it actually contains much higher concentrations of many of the toxins and carcinogens than the inhaled smoke. Benzene, carbon monoxide, nicotine, PAHs and nitrosamines, appear in larger quantities in sidestream smoke. The toxic substances in sidestream smoke are, however, immediately diluted by the surrounding air. Thus the risk to non-smokers is certainly much less than for the smokers. How much less is uncertain and dependent upon the ventilation in the area, how many smokers are present, and how long the non-smoker is exposed.

Health Effects

Surgeon General's report summary

The 1986 Surgeon General's Report drew three major conclusions regarding the health risks of involuntary smoking:

(i) Involuntary smoking is a cause of disease, including lung cancer in otherwise healthy non-smokers;

(ii) Children of parents who smoke compared with the children of nonsmoking parents have an increased frequency of respiratory infections.

(iii) The simple separation of smokers and non-smokers within the same air space may reduce, but does not eliminate, the exposure of non-smokers to environmental tobacco smoke.

A wide range of health effects

The physical effects of exposure to second-hand smoke range from irritation to heart disease or cancer leading to premature death. At the less threatening end of the spectrum, second-hand smoke commonly causes symptoms such as eye irritation, headache, cough, sore throat, nausea and dizziness. A brief exposure to tobacco smoke also impairs lung function slightly, and can raise heart rate and blood pressure. Even fairly small amounts of carbon monoxide from smoke can impair driving ability. Roughly 21% of all Canadians have conditions like asthma, allergies, heart disease, acute respiratory disease or emphysema which are aggravated by exposure to tobacco smoke.

Long-term exposure adds to the risk

The effects of prolonged exposure to second-hand smoke are far more serious than the immediate symptoms. Studies of non-smoking spouses of smokers agree that their risk of developing lung cancer is at least 1.3 times higher than those not exposed to tobacco smoke in the home. An estimated 50 - 60 lung cancer deaths among non-smokers in Canada in 1985 were the result of spousal exposure to tobacco smoke. When other sources of exposure such as the workplace were included, the estimates rose to over 300 lung cancer deaths being caused by involuntary smoking. A similar study in 1986 involving over 7,000 non-smoking women between the ages of 30 and 59 who were married to smokers found that 23 of them had suffered heart attacks. This is three times the heart attack rate for a similar group of women married to non-smokers. It is worth noting that such a correlation does not automatically imply a cause (see Chapter VIII).

Non-smokers suffer at work and in homes

The level of exposure and its duration are key factors in determining the health risks of second-hand smoke to non-smokers. A study of adults who had worked for at least 20 years in offices that allowed smoking showed that they had experienced a loss of lung function equivalent to people who had smoked 10 cigarettes a day for 20 years. Findings such as these have supported the growing numbers of non-smokers objecting to being subjected to second-hand smoke in their workplaces and their homes.

Children are most susceptible to harm

The people who are probably most susceptible, and usually least able to protest, are the children of smokers. A 1990 study estimates that 17% of all lung cancers among non-smokers in the US result from exposure to tobacco smoke during childhood and adolescence. The study found that exposure through childhood and adolescence for 25 or more "smoker-years" doubles the risk of lung cancer. A smoker-year equals one year of living with a smoking parent. Childhood respiratory illnesses also seem to be linked much more strongly to second-hand smoke than to air pollution, even in highly polluted areas. Children of smokers can have a 20% to 80% higher risk of respiratory problems than children not exposed to tobacco smoke.

Regulation of Second-hand Smoke

"No-smoking" measures are stiffening

As an element of indoor air pollution, second-hand smoke is the easiest to control because the sources, the smokers, are obviously identifiable. Measures to control indoor smoke are growing stronger. Earlier policies designating smoking areas in public places are giving way to outright bans on smoking in many buildings as the only way to eliminate the risks to non-smokers. As of January 1990, no smoking is allowed in any federal government building in Canada, including the House of Commons, but excluding members' offices. Airlines in several countries, including Canada, have toughened their smoking restrictions. Air Canada, Canadian International and other Canadian airlines have banned smoking on most flights. US airlines banned smoking on all flights under six hours.

Strong warning messages enforced

The Canadian government, despite strenuous objections from the tobacco industry, has imposed what some are calling the toughest tobacco-warning legislation in the world. After June 1, 1991, all cigarettes and loose tobacco will have to display larger, more prominent warnings which state that tobacco is addictive, that it can harm non-smokers, and that it can cause strokes and lung disease. Cigarette packages will also have to include leaflets with more details and warnings concerning the health effects of smoking.

Restrictions on marketing cigarettes

The tobacco industry is also under seige from ever increasing restrictions on advertising and promotional sponsorship. Already over 20 countries including Canada have passed laws banning all tobacco advertising or sponsorship. Two of the industry's largest markets will likely do the same. The US has banned television advertising since 1971 and now awaits decisions on at least half a dozen bills that would ban or restrict advertising through other media. The European Community intends to harmonize their regulations by 1992. In March of 1990, the European Parliament voted for a total ban on all tobacco advertisements in all types of media.

Bleak outlook in the Third World

However, while tobacco consumption drops in North America and Europe, the number of smokers in many less industrialized countries (the Third World) is very high and increasing. The vast majority of men in China smoke. Sophisticated marketing techniques may increase the proportion of people who smoke in these countries. Few women in Asia and East Europe smoke now; marketing of cigarette brands aimed specifically at women may increase the consumption considerably.

ASBESTOS

Asbestos is a very useful material

Asbestos has properties which makes it useful for many applications. It is fibrous, heat resistant, insoluble in water, and has high tensile strength. Asbestos is used in strengthening cement pipe, in flooring products, automobile brakes, roofing products, packings and gaskets. The best known application is probably the production of heat-resistant textiles used in industry and by firemen.

Canada is a major world producer

Widespread use of asbestos began in the late 1800s, made possible by the discovery of large deposits in USSR, Canada and South Africa. Canadian asbestos is primarily "white" asbestos (chrysotile), while in South Africa, "blue" asbestos (crocidolite) and "brown" asbestos (amosite) are most common. These three major types of asbestos have somewhat different properties. Chrysotile asbestos, which is most abundant, forms flexible, silky fibres that are easy to spin, but it is not resistant to acid. Crocidolite crystallizes as sharp needles and is resistant to acid.

Health Effects

High occupational exposure to asbestos

Reports on respiratory problems among asbestos workers date back to the turn of the century. In those early years of the asbestos industry, illnesses of the respiratory system such as tuberculosis were quite common, and the additional risks of exposure to asbestos were not considered significant. Studies of the health hazards of asbestos in the 1930s and 1940s were not publicized by the industry. In the 1940s, asbestos was widely used as thermal insulation and fire protection for buildings and ships. Pipes and boilers were coated with asbestos, and occupational groups such as shipyard workers, hurrying to replenish and expand the war-time navies, were exposed to very high levels of asbestos dust. Exposure in factories making gas masks was also high.

Asbestos diseases have a latency period

There are three major diseases related to asbestos: asbestosis (a scarring of the lungs), mesothelioma (cancer of the lining of the chest and abdomen), and lung cancer. They all have a latency period of decades. By the 1960s and 1970s the very serious health effects of asbestos exposure became apparent from unexpectedly large numbers of asbestos workers dying of cancers and lung diseases.

Asbestos scars the lungs, causing asbestosis

An estimated 65,000 people in the US are currently suffering from asbestosis. It is a scarring of the lungs caused by prolonged heavy exposure to asbestos dust. Such exposure was possible for asbestos workers decades ago. The symptoms begin with shortness of breath, developing in the advanced stage to near-paralysis of the lungs, making breathing and body movement increasingly difficult. Ultimately, asbestosis leads to heart strain and cardiac failure.

Workers suffer rare cancers much later

Since 1943, it has been established that mesothelioma is caused by asbestos. This is a form of cancer which strikes the lining of the lungs, heart or abdomen. It is very rare among the general population. Asbestos factory workers and shipyard workers experience a very much higher incidence of mesothelioma. Maps of cancer mortality in England show that almost all deaths resulting from mesothelioma occurred in areas close to industrial sites where large amounts of asbestos have been made or used in the last 40 years. The disease takes 30-40 years to appear, but then usually causes death within two years of diagnosis.

Evidence for lung cancer is conclusive

Evidence that asbestos causes lung cancer has been available since 1949 and is now quite conclusive. Many studies have been carried out, some involving groups of more than 10,000 individuals and lasting over 10 years. It is also well established that cigarette smoking combined with asbestos exposure vastly increases the risk of developing lung cancer. While a non-smoking asbestos worker has about five times the risk of dying from lung cancer as an unexposed, non-smoking person does, the risks are over 50 times greater for an asbestos worker who smokes. Like mesothelioma, lung cancer is also usually fatal within two years of diagnosis.

The shape of the fibres is important

It is the physical shape of the asbestos particles that determines the carcinogenicity of asbestos, rather than its chemical nature. The most dangerous fibres seem to be those less than two millionths of a metre (micrometre or um) thick and five to 100 um long. Crocidolite (blue asbestos) fibres seem to be more carcinogenic than those of chrysotile (white asbestos), perhaps because of their greater rigidity.

Workers using asbestos at greatest risk

People who are occupationally involved with asbestos are at greatest risk. Their risk varies with health precautions taken at work, personal habits such as smoking, and the type and size of the asbestos. Due to its uncontrolled use in industrialized countries some decades ago, asbestos was probably the substance which posed the highest occupational risk of cancer. Each year in the US, of the order of 3,000 to 12,000 cancer cases (about one percent of all cases) result from past occupational asbestos exposure. With stricter controls and reduced use of asbestos in products in the US, these death rates are likely to be considerably lower in a decade or two. Nevertheless, over 200,000 workers are expected to die prematurely by the year 2000 from diseases related to exposure to asbestos.

Risk to public is relatively low

There is always some risk of disease from the presence of asbestos, but the risks to the general public are very much lower than to asbestos workers. It is estimated that each year there are about 360 cancer deaths in the US due to non-occupational exposure (333 mesothelioma, 28 lung). Non-occupational exposures are rarely large enough to cause asbestosis. We can come into contact with excess asbestos in several ways: an increase in the general background level of asbestos fibres because of industrialization; fibres introduced into the air from products containing asbestos, such as insulation, brake liners, or tiles, panels and pipes containing asbestos; and ingestion of asbestos in liquids.

Problem of asbestos in schools

A topic of some public concern has been asbestos in the air of schools with asbestos insulation. Some school boards have decided to close their schools and tear out all asbestos. The workers removing the asbestos can be at a high risk during this operation and, if done improperly, considerable amounts of asbestos will be introduced into the neighbourhood. Other school boards have decided that the asbestos is adequately shielded so that negligible amounts will escape; hence the benefit of the removal is not worth the costs and health risks of the removal operation.

Children are at greater risk from exposure

It is especially important to protect younger people from products containing asbestos. Growing children are more likely to be affected by any chemical insult to their bodies, and the life expectancy of youths is much greater than the decades-long latency period for asbestos-induced cancers. Even babies may be exposed to asbestos through the use of talc powder. Some brands of baby powder have been found to contain asbestos, presumably because asbestos can be present as an impurity in the talc used as the raw material.

Presence of asbestos in water

Asbestos filters have been used in the beverage and pharmaceutical industry, resulting from higher than normal asbestos levels in some beverages. Natural water may also contain asbestos from the weathering of rocks. The concentration depends on the geological formations present in the region of the water source, and the degree of disturbance of the soil by activities such as mining and construction. Water can also pick up asbestos contamination when carried through asbestos-cement pipes. Some studies have reported a correlation between the ingestion of asbestos fibres from water and the incidence of cancer. Asbestos workers have excessive rates of stomach cancer, presumably from swallowing asbestos dust breathed in and cleared from the lungs.

Control of Asbestos

More asbestos in air of cities, buildings

Asbestos is a natural mineral; it is therefore present in soil, water and air as a result of natural weathering processes. This "natural background" is believed to be of the order of one fibre in a million cubic centimetres (cc) of air (one cc is about the size of a thimble). In urban areas, the asbestos concentration is about one fibre per 10,000 cc, and in normal indoor air about five times that (one fibre per 2,000 cc). Much higher concentrations occur when products with loose asbestos are in the houses or if asbestos mixing or manufacturing is carried out close-by, because asbestos fibres are carried by wind for considerable distances. The measurement of the concentration of asbestos fibres in air samples requires exceptional care and sophisticated equipment.

Warnings, trade bans lead to reduced use

Since 1979, the use of asbestos in western industrialized countries has decreased dramatically as its harmful potential was recognized. Warning labels were put on products containing asbestos: "Breathing asbestos dust can damage health. Observe safety rules." Some countries like Sweden and Iceland have already banned the use and importation of products containing asbestos. Production of asbestos in Canada has nearly halved and current production is largely exported to Third World countries. The International Agency for Research Against Cancer estimates that there are about 3,000 current uses for processed asbestos, so that many people are still exposed to its dust at work without realising it.

New measures to protect workers

Occupational exposure to asbestos has been reduced in most industrialized countries as labour, government and industry became aware of the seriousness of the asbestos hazard. The maximum allowable limit in many jurisdictions is now two fibres per cc, about 1000 times the level in normal indoor air. The US workplace limit is now 10 times lower than this, 0.2 fibres per cc. Even with this lower limit, it is estimated that asbestos exposure will still cause cancer in six of each thousand workers thus exposed over a working lifetime.

In the US, a ban has been imposed on almost all asbestos-containing products where substitute materials are available. This is to be phased in between 1990 and 1996. The ban includes such products as asbestos-cement pipe, roofing felts, flooring felts, vinyl-asbestos floor tile and asbestos clothing. Starting in 1994, asbestos in automobile brake liners will also be forbidden.

Search for asbestos substitutes

Unfortunately, some asbestos substitutes may themselves pose health hazards, especially if they have fibres small enough to enter the lungs. Substitutes are chosen because their physical properties are similar to asbestos. They may thus present similar hazards, but they have not had the close scrutiny that asbestos has had over the decades. For example, the World Health Organization considers the substitutes rockwool, glasswool and slagwool as "possibly carcinogenic to humans" and ceramic fibres have caused cancer in animals.

Alternatives may also be hazardous

Although less asbestos is used in Western Europe and North America, world production is again increasing because of increased use of asbestos in the USSR, Eastern Europe and many Third World countries. Some of this increase may be due to the moving of asbestos product manufacturing to Third World countries. For instance, brake blocks containing asbestos are now made in Malaysia and Indonesia by an Australian company. In many Third World countries, there is a lack of concern about exposure to asbestos. Exposures from cutting asbestos-cement products with power saws can exceed 100 fibres per cc unless the saws are properly designed and fitted with exhaust suctions and high-efficiency fabric filters for dust capture. Such equipment is rarely available in the Third World.

Asbestos use increasing in Third World

RADON

North Americans became aware of the radioactive gas radon as an indoor air pollutant after an event in Pennsylvania in 1984 sparked widespread media attention. An engineer working at a nuclear power plant was setting off the plant's radiation exposure alarms when he entered the plant in the mornings. The source of the radiation was traced to his home, where radon levels in his basement were found to be many times greater than those usually experienced by uranium miners. Further monitoring of the surrounding area including New Jersey and New York revealed that many houses had very high indoor radon levels. The radon was apparently originating from no other source than the ground upon which the houses were built. A long finger of uranium-containing rock (the "Reading Prong") underlies the houses in that area.

Radon gas from ground around home

Although health officials had been concerned about the risks of radon pollution in houses for some decades, contamination was thought to occur mainly in homes built on lands reclaimed from mining or on top of mill tailings, especially uranium tailings. The situation at Port Hope, Ontario is an example. In 1932, a plant was set up in Port Hope to process ores mined at Port Radium, Northwest Territories. Since then, radioactive waste material has been stored in disposal areas around Port Hope. In the early years, disposal practices were quite lax

More radon in homes built on mill tailings

and it became clear by the mid-1970s that contaminated waste around and under houses was causing higher than acceptable radon levels in many buildings in Port Hope. The federal government spent millions of dollars between 1976 and 1981 trying to correct the problem by removing contaminated materials from landfills and buildings. Their efforts affected about 400 homes and buildings in Port Hope.

Concern over radon increasing

High radon levels in the houses on the Reading Prong not only captured public attention, they also contributed to a growing number of cases which made it clear that the potential for indoor radon pollution was far more widespread than previously thought. Many areas in both the US and Canada generate significant quantities of radon naturally. This realisation heightened debate about suggested links between indoor radon levels and lung cancer and consequently about the extent of the risk indoor radon actually poses to the public.

Radon forms through decay of uranium

When radon was first discovered in 1900, it was referred to as "radium emanation" until it was recognised as a distinct element. Radon is an odourless gas formed in the ground as one of a series of radioactive substances produced by the natural decay of a type (or isotope) of uranium known as uranium-238 (see Appendix B). The immediate precursor to radon in this radioactive decay chain is radium-226. Uranium-238 exists naturally in trace amounts in most soil and rock, particularly granite (see Chapter VII). Thus, radium and radon also exist naturally, mostly in rocky ground.

Many sources of radon exposure

Soil is by far the largest natural source of radon, but oceans, groundwater, uranium mill tailings, phosphate minerals and coal residues are included among the known additional sources. Terrestrial radon is, in turn, a part of the broader category of terrestrial, cosmic and atmospheric radiation totalling roughly 1.85 millisieverts (mSv) per year (see Appendix B for an explanation of units) to which we are all exposed. Sources of radiation from human activity total about 0.34 mSv per year from such things as medical exposure from X-rays, fallout from nuclear weapons tests, nuclear power and other sources like television sets.

Effective average doses in mSv per year (Appendix B)

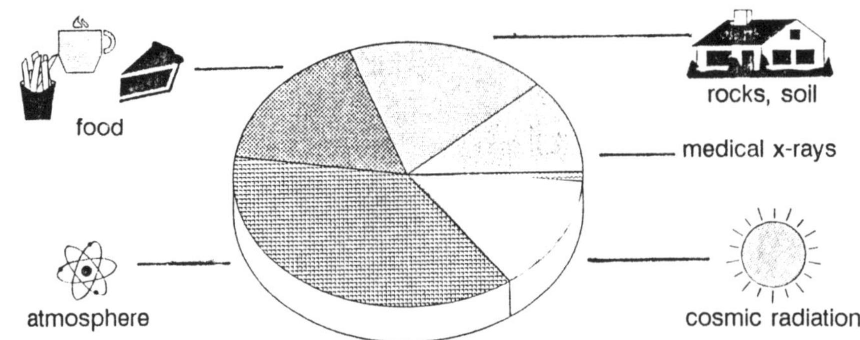

Sources of Radiation Exposure

Radon gas is dense, chemically unreactive

Radon is a very heavy gas, about 7.7 times heavier than normal air. It is also an inert gas, and does not react chemically with any materials. It can seep up out of the ground into water or air. Normally radon escapes to the air where it is dispersed by wind, but underneath a building it may seep in through cracks in the basement and become trapped in the building. Once inside, it tends to rest in the lower levels of a building because of its density, accumulating from the ground up.

Radon level in ground is most important

Early research on indoor air pollution assumed that the "leak-proofing" of buildings to conserve energy was the major cause in the build-up of radon, because the sealing up of a house would lower the ventilation rate in that house. Closer examination of houses with high radon levels found little or no relationship between the airtightness of a house and its radon level. Apparently, the differences in radon levels between various houses are due mainly to differences in the strength of the radon source underground. In addition, some researchers believe that houses actually draw radon from the ground because of differences in pressure between the inside and outside of a house caused by wind and temperature effects. It seems that, just as a fire draws air into a fireplace and up the chimney, a heated house can draw air in from the ground around the basement.

Pressure can flush radon out

Radon seems to be difficult to remove. Increasing ventilation is not a very practical remedy for most radon problems. A more effective solution seems to be to reverse the pressure difference that draws the radon into the house in the first place. One way to do this is to drive pipes through the basement floor and use small fans to push air from the underlying ground to the outdoors.

Health Effects

Decay products of radon are harmful

The most important isotope of radon in terms of health effects is radon-222. It decays with a half-life of four days, producing a sequence of break-down products ("daughters") all with much shorter half-lives than radon. All of the daughters produced by radon are more harmful than radon itself, but the most important are two isotopes of polonium. These substances can be present in fine aerosols and are chemically active. They can attach to airborne particles, walls, and, if inhaled, to the lining of the lung. Once inside the lung, they are likely to continue their radioactive decay, irradiating the surrounding lung tissue in the process.

Lung tissue harmed by alpha radiation

The most damaging radiation dose to the lung comes from alpha particles (see Appendix B) emitted when the polonium isotopes decay. Alpha particles can barely penetrate the thickness of a piece of paper (about 70 um) or the dead layer of skin on a finger, so that they are not very harmful unless they are emitted inside the body after being ingested or inhaled. Unfortunately, the polonium daughters of radon provide an efficient system for irradiating the lung cells crucial to the formation of cancer. The layer of lung tissue that contains cells which can generate cancer is only about 40 um thick - right in the penetration range of the alpha particles.

Uranium miners have had more lung cancer

The link between radon and lung cancer among uranium miners has been recognised since the 1930s. In the 1950s, the first detailed studies of radon effects on miners were conducted in the US and by 1955, the International

Commission on Radiological Protection (ICRP) had set a health standard of ten picocouries (pCi) of radon per litre of air.

Linear response, cumulative effect

As researchers have turned to the task of estimating the risks of developing lung cancer from exposure to household radon, they have relied heavily on what is now a large body of data from underground miners in the US, Canada, Sweden, Czechoslovakia and many other countries. The ICRP extrapolated the data from the miners' occupational exposure to radon to estimate the risks posed by radon in homes. These risk estimates assume that there is no threshold radiation dose below which the radiation cannot cause adverse health effects (see Chapter VIII). Another assumption made by the ICRP is that the effects of radiation on humans varies with the total dose, independent of the rate at which the dose is delivered. Thus it is assumed that low exposure to radiation over a long period of time is equivalent to a higher exposure level over a shorter period.

Lung cancer rate estimates by EPA in USA

Some countries have concluded that radon is second only to smoking as a cause of lung cancer. Based on surveys of indoor radon levels in 17 states in the US, the EPA estimates that inhalation of radon is probably responsible for 5,000 lung cancer deaths among non-smokers every year. Incorporating evidence that smoking increases the risks by up to 15 times, the EPA adds another 15,000 yearly deaths from the combined effects of smoking and radon.

US limits on radon lower than Canadian

Accordingly, the US has set an "action" level for homes and buildings at four picocuries of radon per litre of air believing that a lifetime exposure (70 years) to this level of radon gives an individual a one to five percent chance of dying from lung cancer. This is considerably lower than the 20 pCi per litre (800 Becquerel per cubic metre) guideline which exists in Canada.

Risk may be overestimated by EPA

Some scientists feel that the health risks arising from indoor radon have been exaggerated. Most researchers agree that indoor radon can be dangerous at levels in the 10 to 20 picocurie range. Exposure to radon in this range over a lifetime can approach the exposures that miners received when increased lung cancer rates were observed. There is however some feeling among scientists that the EPA set its action level too low.

No excess lung cancer in high-radon regions

They note that in many of the studies of uranium miners, most of the men examined were smokers and argue that any estimates of lung cancer deaths from indoor radon based on these studies are only relevant to smokers. Some studies which have examined areas with very high natural emissions of radon have not found the predicted excess lung cancer rates. Such results do not imply that radon does not cause cancer, but their proponents challenge the "no-threshold theory" of cancer risk from radiation. It might be that radiation can even have a beneficial effect on the body at levels below those at which harmful effects are produced (see Chapter VIII).

Radon poses a relatively great risk

Radon is an unusual pollutant, no one is responsible for it, yet everyone is exposed to it. The risk of developing cancer in a lifetime from radon may be of the order of one percent. This is large compared to the risks frpm most other chemicals in the environment. For instance, some regulated carcinogens like benzene or various pesticides carry a lifetime risk of no more than 0.001% for the average person. It is however small compared with some of the risks of death that we seem to accept from voluntary actions. The risk of death from driving is two percent while smoking carries a risk of premature death as high as 25%. People made the choice a long time ago to build solid structures to live in for comfort, safety and, ironically, health. Without knowing it, people have always lived with some level of health risk from indoor radon. Now that the problem has been recognized, the task is to find out the magnitudes of the risks and then to set practical approaches to managing them.

CONCLUDING REMARKS

The problems are localized

This chapter deals with very localized environmental pollution, as compared to the regional problems discussed in Chapter I and the global ones in Chapter II. Thus regulation can be more localized and remedial action can be rapid.

The problems are greater in cold climates

Many of the problems discussed in this chapter are more relevant to countries with a cold climate. We build more airtight houses and spend a larger fraction of our time indoors than people living in milder climates. The problems of indoor air pollution may actually get worse if we continue to push for energy conservation through more energy-efficient housing, while at the same time not taking care to ensure proper ventilation.

Real cause of sick buildings not clear

There is now some acceptance that sick-building syndrome is an actual physical condition, and not simply a psychological response to a hostile world containing synthetic chemicals. Allergic attacks with symptoms such as headaches, depression, fatigue or nausea may be caused by a depressed immune system due to hypersensitivity to the synthetic chemicals contained in a variety of common products. This ailment is known by some as "environmental illness". Some immunologists nevertheless claim that such "multiple chemical hypersensitivity . . . constitutes a belief and not a disease". Some common allergic reactions such as hay fever are certainly very real. Such allergies are often due to spores from moulds and fungi and not chemicals. Because of such controversies, the real cause of sick buildings is not clear. There seems to be little doubt that good ventilation is a wise step towards preventing or curing this condition. It is especially important in a new building, since the building and its furnishings give off more chemicals when new.

Involuntary smoking risk not acceptable

In the last decade, we have made substantial progress in Canada in reducing second-hand, as well as first-hand, smoke. Opinion about the acceptability of smoking in public seemed to change when good evidence became available about the high risks of second-hand smoke. Smoking is a voluntary activity, and smokers may be ready to accept the risk. Exposure to second-hand smoke is however usually involuntary, and being exposed to such risks is much less acceptable (see Chapter VIII).

Asbestos risk small from home exposure

Before the 1970s, the real risk to health from asbestos was very high for those occupationally exposed. The risk from environmental exposure in buildings is however very small relative to risks from radon or second-hand smoke. Yet there has been great concern about asbestos insulation in schools. This is probably because children are forced to go to these schools. The risk is involuntary and it is to the very young.

Extreme exposure in shipyards

The perception of risk posed by asbestos is probably also overestimated because people forget to adjust for the level of exposure. Asbestos workers were exposed to massive amounts of the substance, especially during the war years when naval shipyards were working frantically. If properly installed, the presence of asbestos insulation in a building may not result in any increase in the asbestos levels in the air of the building above the natural background.

Risks from radon not well known

The risk from exposure to radon is still under debate. It may be that the risk is overestimated by 30% using data on miners because of two factors not previously considered: the breathing rate of the people involved, and the size of dust particles. Some scientists claim that it could be even lower, as there is evidence that the risk is not a linear function of dose. Even if the risk of radon is currently overestimated, it is nevertheless a major cause of lung cancer in a country such as Canada.

Why radon risk may be more acceptable

In spite of the relatively high risk from radon, people do not seem overly alarmed about it. Radon is the focus of much less concern in the media than asbestos, even though it presents a much greater risk. Radon occurs naturally, as does asbestos, but radon is not placed in the house by anyone the way asbestos is. There was, however, considerable alarm about radon levels in homes in Port Hope; perhaps because in that case soil contaminated by an industry was found to be responsible.

Recommended Reading

"Radon Tagged as Cancer Hazard by Most Studies, Researches", D.J. Hanson. *Chemical and Engineering News*, pp 7-13, 6 February, 1989.

"Asbestos: Medical and Legal Aspects", 2nd ed, B.I. Castleman. Prentice-Hall, Clifton NJ, 1986.

"The Health Consequences of Involuntary Smoking", a report of the Surgeon General. US Department of Health and Human Services, 1986.

"Radon: a Homeowner's Guide to Detection and Control", B. Cohen. Consumer Reports Books, Mt Vernon NY, 1986.

"Asbestos: the Fibre that's Panicking America", P.S. Zurer. *Chemical and Engineering News*, pp 28-41, 4 March, 1985.

CHAPTER IV
POLLUTION OF WATER

Possibly nothing is more appealing on a blistering-hot summer day than the thought of languishing on a seaside beach or plunging into the cool, clear water of a local river or lake. But for more and more people in Canada and the rest of the world, this wish cannot be a reality. Whether due to pollution from sewage, industrial effluent or agricultural run-off, the water near many people's homes is not fully safe even for a quick dip.

Few natural waters are safe for swimming

Maybe an ice-cold glass of water could help to cool things off - but, then again, maybe not. Traces of pesticides and industrial chemicals have found their way into some domestic water supplies. Everyday activities like washing the dishes, taking a shower or flushing the toilet, which most of us take for granted, contribute to the burden placed on our water.

Everyday activities require water

Yet we in Canada are relatively fortunate when it comes to the quality of our water. Many people in the world have difficulty getting enough water even for essential uses. Billions of people are at risk from serious waterborne diseases such as cholera and typhoid. Millions of these people, mostly young children, die of such diseases every year. These diseases are the result of biological rather than chemical contamination; hence the pollution problems responsible for them are outside the scope of this book. The contribution of chlorine in combating biological contamination and assuring a safe water supply is discussed in some detail in Chapter V.

Billions of people have no clean water

In Canada, we have essentially eliminated waterborne diseases, but the stresses from human activities have put the health of one of the world's most important freshwater ecosystems in jeopardy. The Great Lakes basin spans the provinces of Ontario and Quebec and six states in the US. It is home to more than 40 million people, including over half of Canada's population. Three-quarters of Canada's industrial activities and roughly half the dollar value of Canadian agricultural production are contained within the Great Lakes region, where gross annual production on both sides of the border exceeds one trillion US dollars.

Great Lakes ecosystem under threat

The Great Lakes have paid the price for their utility to our industrialized society. They endure pollution from municipal and industrial sewage, leaking landfills and toxic waste dumps. Run-off from agricultural and urban areas end up in the lakes carrying wastes like pesticides, fertilizers, animal wastes and lead from car exhausts. Pollution can travel long distances, through groundwater and the air, so that sources can be difficult to identify. For example, airborne fallout, possibly from activities thousands of kilometres away, seems responsible for most of the pollution of Lake Superior.

Pollution sources varied and numerous

Unfortunately, the predicament of the Great Lakes is not unique. Indeed, they provide a good example of the various kinds of pollutants that are affecting many

Many bodies of water polluted

waters around the world. To this continuous pollution burden there is added the effects of tragedies such as the Exxon Valdez oil spill of 1989, and the oil releases into the Persian Gulf in 1991. It is evident that human activities are leaving a legacy of water pollution.

AVAILABILITY OF WATER

The earth's water cycles constantly

From a Canadian perspective, it is easy to believe that supplies of freshwater are limitless. Canada contains one-seventh of the world's lakes and rivers - enough water to flood the entire country to a depth of over two metres. There is a sense of timelessness, immortality, about a river flowing endlessly to the sea. And if a well runs dry, then dig another. The supply of water is, in a sense, endless because it is continuously being recirculated from the oceans to the atmosphere through evaporation, falling to the ground as precipitation, then flowing back to the sea in rivers or groundwater.

Volumes of reservoirs in km^3 of water

Flows in thousands of km^3 per year

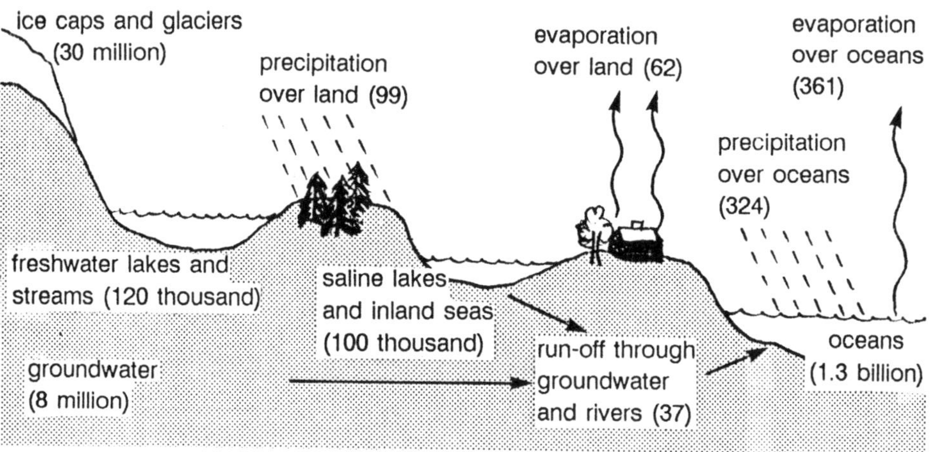

The Water Cycle

Most of earth's water is out of our reach

Global reserves of freshwater certainly exceed all conceivable needs of the earth's human population. Roughly 120,000 cubic kilometres of freshwater exists on earth, enough water to fill the Mediterranean Sea ten times over. But of that total, three-quarters is out of reach, bound up in glaciers and polar ice. Only 0.03% of the remaining one-quarter exists in the form of lakes, rivers and water vapour in the atmosphere. The rest lies underground, percolating through the soil or pooled in underground aquifers as groundwater.

Water is a rare commodity in many regions

Unfortunately, water is not evenly distributed on the planet. The same climatic processes that bequeath such a wealth of water to Canada produce vast deserts and semi-arid regions in the world. In such areas, obtaining the two litres of water a day necessary for human survival can be an all-consuming task. In stark contrast, the average Canadian uses about 300 litres of water every day, almost half of which is literally flushed down the toilet.

UN target is clean water for all

It is a sad reality that in many parts of the world, mainly in the Third World, whatever water is available is so polluted that it is extremely dangerous to use. In India, nearly 70% of the available water is polluted and waterborne diseases like cholera and typhoid account for 80% of all illnesses. Statistics like these spurred the United Nations to launch a campaign in 1980 to provide clean water and sanitation for all by 1990.

Water problem persists, even with progress

However, while in this decade 542 million more people gained access to clean water and 357 million to sanitation, the rise in world population more than offset the gains. The World Health Organization (WHO) estimates that there are now 28 million more people worldwide with no access to clean water, bringing the total to 1.23 billion people. As well, those without proper sanitation now stand at 1.8 billion, 213 million more people than in 1980.

Sewage increases with population

As a rule, countries in the Third World lack adequate sewage treatment and waste-disposal systems. In the case of the Ganges river in India, whose waters hold great religious significance to the millions of Hindus who bathe in it, the problem of inadequate sanitation has been exacerbated by years of population growth, urbanization and industrialization. In developed countries, the threat posed by biological pollution of drinking water has been largely eliminated. This is discussed more thoroughly in Chapter V. Ironically, in many of the developing countries, biological pollution has now often been replaced by the threat of chemical water pollution.

Faecal waste fouls beaches

We may be free from biological contamination of our drinking water, but the closure of beaches every summer serves to remind us that faecal wastes are still a water pollution problem. In areas served by wastewater treatment, most such pollution results from animal excrement carried in from streams and storm sewers. During heavy rainstorms, faecal wastes and other pollutants such as fertilizers are washed from the streets and lawns into storm sewers which then overflow into rivers.

Canada slow to build sewage treatment plants

Only about 57% of Canada's population is served by wastewater treatment plants, compared to 75% of all Americans, 86% of Germans, and 99% of Swedes. The province of Quebec has the worst record in Canada for sewage treatment; only 23% of its city residents are served by sewage treatment which does more than just screen out solids. The city of Montreal opened its first sewage treatment plant as recently as 1987. Before that, the city sent roughly 600,000 cubic metres per day of raw sewage into the St. Lawrence River.

SURFACE WATERS

Myth of endless water supply causes abuse

The bodies of water on this planet serve the needs of humans and all other living things in countless ways. They provide us with food and drinking water; they are our transportation routes and playgrounds; they are used in industry and agriculture. Paradoxically, they are also perceived to be first-rate garbage dumps. Dumping of garbage and dangerous wastes into our lakes and rivers continues continues largely because of a long-held belief that freshwater exists in boundless quantities and has the inherent ability to absorb wastes and wash them out to seas where these wastes will, in some mysterious way, disappear.

Many uses of water can lead to pollution, deficiencies, and conflicts

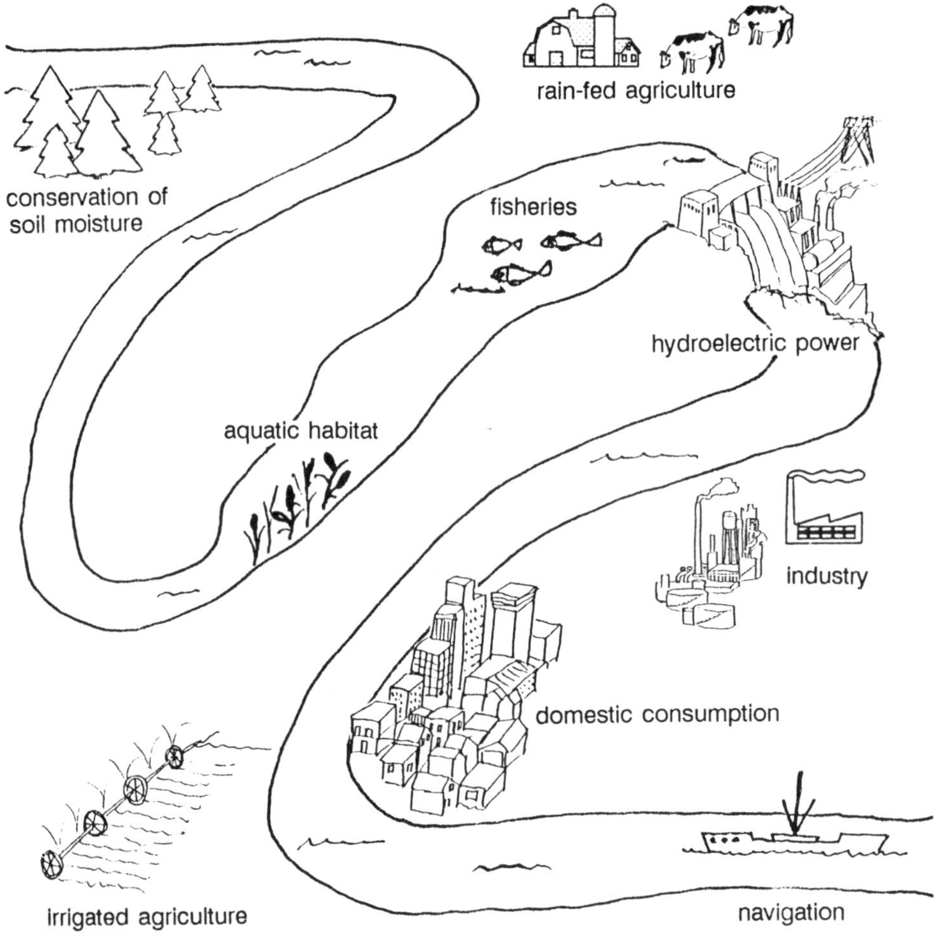

Multiple Uses for Freshwater

Water cleans itself only to a limited extent

Only to an extent is this belief correct; water can "clean itself". Aquatic bacteria can break down organic materials such as human and animal wastes. This process has limits, often because the oxygen in the water is depleted. Moreover, many of the toxic chemicals dumped into waters worldwide can not be removed by these natural means, and they become persistent pollutants.

Chemical Pollution

Chemicals from industry can be persistent

Chemicals are a vital part of our industrial society - over 100,000 commercial chemicals are involved in our economy. The industrial and agricultural processes which enhance our standard of living also create countless by-products, some of which are both toxic and very long-lived. The persistence and mobility of some of these chemicals is illustrated by a case involving the Dene people of the Northwest Territories.

Pollution transported long distances

The liver of the losh, a fish that lives in the Mackenzie River, is considered a delicacy by the Dene. In 1987, they began to notice a colour change in the livers of the losh that they were catching. Federal scientists investigated the problem and discovered that the pesticide toxaphene was accumulating in the livers of the losh. Farmers in the southern US used to spray toxaphene on cotton fields, before it was banned a decade ago. Not only was this chemical present ten years after its use, but it had travelled thousands of kilometres through the air, finally ending up in the livers of fish in the Northwest Territories in Canada.

Pollutants in air can end up in water

This is just one example of the impacts of long range transport of airborne pollutants. Vast quantities of natural and synthetic chemicals are emitted into the atmosphere on a daily basis. These substances are dispersed throughout the globe by air currents; they know no international boundaries. Slowly the pollutants carried on the winds come back to earth as precipitation. Acid rain, as discussed in Chapter I, is the best known example of this phenomenon. When these pollutants fall to earth, they can contaminate soils, and ultimately also water. The environmental and health impacts of this kind of pollution are difficult to evaluate because they can occur over varying time frames and over vast areas.

Pollution most severe in populated regions

In more heavily populated areas, the contribution of atmospheric transport is far outweighed by the direct impacts of pollution from municipal and industrial sewage and agricultural run-off. The Fraser River in British Columbia and the St. Lawrence River continue to be seriously contaminated by chemicals. Traces of over 800 chemicals have been found in the Great Lakes. These include persistent compounds such as methylmercury, DDT, dioxins, and PCBs (see Chapters V, VI and VII).

Factory leaks add to water pollution

In addition to constant pressure on waterways from toxic chemicals contained in wastewater effluent, leaking industrial sewage pipes and storage tanks often contribute unfiltered chemical waste to surface waters. A well-publicized example of this occurred in 1985-86 in the St. Clair River, which flows between Ontario and Michigan. Over several months, thousands of litres of chemicals had leaked from a chemical plant's sewage pipe into the river. These chemicals were mostly chlorinated compounds which are not very soluble in water and which are not broken down biologically to any appreciable extent. As a result, they formed a dark brown "blob" on the river bottom near the sewer leak. This contamination was eventually removed.

Households are sources of toxic wastes

While industry is often blamed for chemical water pollution, households also contribute to the problem. Toxic wastes such as pesticides, cleaners, automotive oil, paints, batteries, and antifreeze all find their way to municipal sewage and waste sites. In communities with well-regulated industries, homes may be greater contributors to toxic pollution than industry.

Species can disappear due to pollution

While the health effects of low levels of toxic chemicals upon humans are still being debated, the impacts of chemical pollution on animal life have become clear. Widespread birth deformities have been discovered among sensitive species of birds like terns, herring gulls and osprey, while disease and reproductive failure are taking an unusually high toll on fish-eating mammals like mink and otters. In Lake Ontario, seven of the ten most highly valued species of fish have almost completely disappeared. This may be due more to the change in the ecology of the Great Lakes from eutrophication than to the presence of toxic pollutants.

Lake Ontario clean-up project has started

These problems have not gone unrecognised. The latest attempt to address the problem focuses on Lake Ontario, which suffers most severely from chemical pollution. Called the Lake Ontario Toxics Management Plan, this effort involves the governments of Canada, the United States, New York State and Ontario. Its goal is to restore Lake Ontario to a state where it can provide safe drinking water and fish. It remains to be seen whether the political and financial will required to succeed can be maintained.

Accidents add to chronic pollution

The pollution of water by toxic industrial chemicals can be caused by slow, steady influxes from sources such as leaking landfills and inadequately treated wastewater. It can also result from accidents, such as the one which devastated aquatic life in the Rhine River in 1986. The Rhine runs from Austria and Switzerland to the North Sea in the Netherlands. It is a major industrial artery. It is also a principal water source for over eight million people, as well as the principal waste depository for municipalities and industries alike. The ecosystem of the Rhine has been stressed for centuries, but increasingly strict regulation of industrial and municipal emissions was steadily improving the quality of the water in the Rhine.

Rhine River polluted from accident

In November 1986, an explosion occurred at a chemical warehouse near Basel, Switzerland. About 30 tonnes of agricultural chemicals and solvents escaped into the Rhine, together with 200 kg of mercury. The accident killed aquatic animals and plants for about 250 km downstream. Even in a river which is continuously flushing itself out, it takes years for the water to recover from such a catastrophe.

Nutrient Pollution

Algal growth threatens other aquatic life

The consequences of pollution from municipal and industrial sewage gained widespread public attention in Canada and the US in the mid-1960s through a crisis involving Lake Erie. Massive algal growth was causing the death of many of the fish and other aquatic life in the lake, leaving stinking, slimy streaks of green scum along its shore. The cause for this accelerated eutrophication was traced to phosphates which were entering the lake with domestic sewage and agricultural run-off.

Eutrophication process accelerated

Eutrophication is due to the primary biological activity which occurs when sufficient amounts of all nutrients are present. It is actually a natural process that converts lakes to marshes and eventually to forests. Natural eutrophication usually takes thousands of years, but it can be accelerated wildly by human contribution of nutrients. Between 1900 and 1970, Lake Erie "aged" the equivalent of 10,000 years due to the availability of unnaturally high amounts of nutrients, especially phosphates.

Identification of controlling nutrient

Other chemicals besides phosphates can cause eutrophication; nitrates and potassium are also important primary nutrients for aquatic life. The availability of one of these, or even a minor essential element, can control the rate of eutrophication. It is important to find out which nutrient is controlling the growth of algae and other aquatic plants; reducing the level of a non-controlling nutrient will do little to reverse the damage. The algal blooms which characterize eutrophication are evident not only in lakes, but also in coastal marine waters. However, while the cause of eutrophication in lakes is usually excess phosphorus, in marine waters it is often caused by extra inputs of nitrates.

Phosphates in Eutrophication

Interference with phosphorus cycle

Phosphates themselves are not toxic to life - in fact, quite the opposite. Phosphorus is essential to life; it is needed in many cell functions including growth, transfer of energy and reproduction. Like other essential elements it follows a natural cycle in the environment. The extra inputs of phosphates to waterways from human activity throw the natural phosphorus cycle off balance, providing aquatic plants with unusually large amounts of phosphorus.

Algal blooms rob water of oxygen

Excessive quantities of phosphates lead to explosive growth of aquatic plants, particularly algae. Algal blooms cause several problems for a lake's ecosystem. For one thing, they can produce toxins that cause illness or even death among birds and mammals. They can also kill submerged plants by blocking essential light. When the algae die, they fall to the bottom where bacteria "eat" them and all other dead plant and animal matter. In the process, the bacteria use up enormous amounts of oxygen as they decompose the material, setting up a vicious cycle that slowly suffocates fish and bottom dwellers such as crayfish.

Wastewater main source of phosphorus

Although phosphates can enter waterways from urban and agricultural run-off and shoreline erosion, the major route is via municipal and industrial wastewater. Industrial effluents and cooling waters can all contain phosphates, but detergents have, at least in the past, been major contributors. Most household laundry detergents contained polyphosphates, which assisted the cleaning action in several ways. It was estimated that, prior to 1970, 30-70% of the phosphate input to water in North America was due to phosphate detergents. These have been widely used since 1946.

Action taken on pollution of Great Lakes

Recognition of the severity of phosphate pollution in the Great Lakes forced the governments of both nations into action. In 1972, the Canadian and US governments signed the Great Lakes Water Quality Agreement. It required that the phosphorus content of effluents from all large municipal waste treatment plants discharging into Lakes Erie and Ontario and the international section of the St. Lawrence River be limited to one milligram per litre (mg/L). To achieve this, the governments of both countries focused on two major tasks: the slow, expensive process of installing or upgrading sewage treatment plants to enable them to remove phosphorus from water before returning it to the lakes; and the removal of phosphate detergents.

Regulation of phosphorus in detergents

The phosphate in detergents was dealt with first, since a reduction seemed to be the fastest, most economical way to significantly reduce the phosphate inflow to the lakes. By 1972, Canada had restricted the maximum allowable phosphorus content for detergents to five percent by weight. The government also allowed any laundry detergent with less than one percent phosphate to declare itself "phosphate free" - a privilege that has been exercised more often since the revival of environmental awareness in the late 1980s. In the US, the regulation of phosphate detergents was left to the governments of the Great Lakes States. By the early 1980s, most of these states had passed such regulations, with Michigan and Minnesota banning them outright. Canadian phosphate regulations are limited in scope. They do not apply to dishwashing liquids or special purpose compounds, all of which can contain up to 45% phosphate. The matter is further complicated because the phosphate replacements themselves have become controversial.

Phosphate replacements controversial

One of the prime contenders to replace phosphates in detergents was nitrilotriacetic acid (NTA). It has, however, been shown to be a suspected carcinogen; NTA was found to cause urinary tract tumours in laboratory rats and mice given large doses over their lifetimes. The amount of NTA likely to be present in drinking water would be so low that no significant risk would be involved. New York State has banned NTA; Canada has approved it for use in detergents, but drinking water levels cannot exceed 0.050 mg/L. The maximum risk of developing cancer over a lifetime at these limits is estimated at about one in 2,000,000 (see Chapter VIII).

Progress in reduction of phosphates

Overall, efforts to stem the flow of phosphorus into the Great Lakes have been successful. In addition to the reductions in phosphates from detergents, sewage treatment plants which can remove the phosphates have been built on both sides of the border. By 1980, the municipal phosphorus loadings to Lakes Erie and Ontario had fallen to about 25% of their 1972 levels. Control of point sources of phosphorus has had some immediate results, notably in Lake Huron's Saginaw Bay and Lake Ontario's Bay of Quinte. These repairs to the water quality in the Great Lakes have been expensive. Since 1971, federal, provincial and state governments have spent at least nine billion dollars on sewage treatment for the lakes.

Case of Erie is worst, but it is improving

The five lakes in the Great Lakes system function as one single ecosystem, so that reducing pollutants in some places but not others can make it difficult to really solve the problem. The problem has not been as significant in the deeper lakes such as Lake Ontario. Eutrophication in Lake Erie has been particularly severe since it is shallow, and since wastewater from several major industrial cities flows into it. The response in Lake Erie as a whole has been slow. The lake is nevertheless improving; oxygen levels in the water have increased and the weeds and algae are not as prolific as they once were. Lakes such as Erie are complex ecological systems, and have many pollution stresses upon them. Recovery of the water quality is bound to be a slow process.

Nitrates in Eutrophication

Nitrates blamed for marine eutrophication

Marine eutrophication, particularly in the Baltic Sea, has become a serious problem. Nitrates entering with run-off from fertilized fields are believed to be primarily responsible. All of the familiar symptoms are present: algal blooms, decreased water transparency. About 100,000 square kilometres of the Baltic Sea are now suffering from oxygen deficiency; as a result, lobsters and some species of fish are dying.

North Sea algal blooms affect fish farms

Accounts of the effects of eutrophication are not limited to the Baltic Sea. They are coming in increasing numbers from bays and estuaries along the coastlines of Japan, China, North and South America, Africa and the rim of the Mediterranean Sea. Algal blooms in the North Sea have caused the oxygen levels to fall dangerously low four times since 1980, with serious consequences to marine life. In the summer of 1988, fertilizer run-off was blamed for the algal blooms that nearly ruined the salmon and trout farms off the coast of Norway. Local fish farmers lost roughly 200 million US dollars.

Run-off from fields unabated

Even facing such consequences, European countries have continued to dump more than 1.5 million tonnes of nitrogen into the North Sea every year. Two-thirds of this nitrogen load comes from rivers carrying agricultural run-off. Dramatic growth in human population along coastal areas and associated pressures to increase agricultural yields mean that many of these areas are among the most heavily fertilized environments on earth. Unless farming practices change dramatically, there is reason to believe that fertilizer nitrogen will continue to run off into coastal waters at high rates.

Nitrogen compounds reach lakes

All living things require nitrogen to make proteins for healthy growth - plants are no exception. Although nitrogen composes about 78% of the atmosphere, plants cannot use it directly. Instead, they take it up through their roots in the form of ammonium or nitrate ions. Both of these forms of nitrogen are soluble in water and can leach through soils into groundwater or run off into surface water. Animal wastes are often rich in ammonium ions, which can easily be leached out by rain. These nitrogen compounds ultimately reach lakes, rivers or the sea.

Fertilizer use introduces nitrogen

For centuries farmers have tried to maintain nitrogen levels in their soils by using manure and rotating crops, but these methods were land and labour intensive. Chemical fertilizers enabled farmers to focus on growing more lucrative crops and achieving greatly enhanced yields from the same amount of land to meet the demands of a growing population. In 1950, farmers worldwide used 14 million tonnes of fertilizer. By 1985, this figure had shot up to 125 million tonnes. It is now often taken for granted that conventional agriculture cannot do without nitrogen fertilizers.

Other sources of nitrogen pollution

Although run-off from nitrogen fertilizers is usually cited as the largest source of nitrogen pollution, it is not the only source. Nitrogen oxides produced by automobile combustion are usually associated with air pollution, but they contribute to water pollution as well when they come back to the earth dissolved in rain (see Chapter I). The problem is further complicated when forests and wetlands are converted into fields. The removal of trees can cause field run-off to increase by two to eight times. Nitrate pollution increases in heavily cultivated areas.

Freshwater and seawater are both influenced

Nitrates are being found in increasing quantities in coastal seawater and freshwater. In seawater, they can cause algal blooms. In freshwater, nitrates are not usually responsible for eutrophication. They are however implicated in issues of human health.

Nitrates in Drinking Water

Babies at risk from nitrate pollution

In very young babies, high levels of nitrates in drinking water can have serious consequences. Nitrates themselves are not lethal, but once inside an infant's gut, bacteria can reduce nitrate to nitrite. Nitrite alters the haemoglobin in blood so that it can no longer transport oxygen to body tissues. As the child is effectively suffocated from the inside, its lips and then body begin to turn blue. The WHO reports that between 1945 and 1986, there were 2,000 cases of nitrite poisoning, in which 160 babies died. In all cases the babies had drunk water with over 25 mg/L of nitrate. It should be noted that this risk is quite insignificant as compared to that of waterborne infectious diseases.

Cancers linked to nitrates in drinking water

The second hazard associated with nitrite is that of conversion in the body to form nitrosamines. Animal studies have shown that there are at least 100 known carcinogenic nitrosamines. Some researchers link nitrosamines to human cancers of the stomach and windpipe. In the early 1980s, a Chinese study reported that areas with high rates of stomach cancer also had higher than average nitrate levels in drinking water. However, a British study reported in 1984 that in areas with high nitrate levels, stomach cancer was actually becoming less common.

Europe regulates water treatment

In an effort to reduce the level of nitrates in water, the European Parliament has recently passed legislation limiting nitrate concentrations in treated municipal wastewater to 10 mg/L. In all surface water and groundwater it will be limited to 50 mg/L. Nitrate removal from wastewater is a very expensive and difficult proposition. Furthermore, legislation of this kind cannot tackle run-off that enters waters directly. This is one reason why many people feel that the pollution should be prevented at its source. This goal can perhaps only be achieved by "re-revolutionizing" agricultural practices (see also Chapter VI).

Benefits of organic farming questioned

Organic farming has been suggested as a solution. It uses only "natural" pesticides and fertilizers such as manure or compost. However wholesome its image has become, this kind of farming could actually increase the amounts of nitrates entering waterways. When applied with proper techniques, manure can produce high yields as effectively as chemical fertilizers. Unfortunately, some research has shown that it also contributes as much as 100 kilograms per hectare more nitrate than artificial fertilizers.

Other traditional methods that are being reconsidered are those of crop rotation or growing two different kinds of crops together - for instance legumes (such as soybeans), grown in combination with cash crops. Legumes, with the help of certain soil bacteria called rhizobia, are able to capture atmospheric nitrogen directly and convert it into a rough equivalent of nitrate within the roots of the plant.

Crop mixing utilizes legume nitrogen

The rhizobia themselves have now become the focus of yet another possible alternative to chemical fertilizers. The rhizobia infect the roots of legumes causing the formation of nodules in which the "natural" fertilizer is made. Many researchers are trying to develop strains of rhizobia that are capable of nodulating non-legume crops like corn, rice and wheat plants. This is a very complicated task, and work is progressing slowly. As well, since the development of these strains would involve genetic engineering, it is likely that this solution will raise its own set of problems in the public forum.

Research in genetic engineering

GROUNDWATER

Water pollution is most obvious to us when it occurs in surface waters. However most of the world's available freshwater reserves exist as groundwater lying below the soil. It is invisible, it is a significant supply of drinking water, and it, too, is being contaminated. Large numbers of people in the world rely on groundwater from wells or springs as their source of drinking water.

More ground-water than surface water

Twenty-six percent of all Canadians rely on groundwater for drinking; in Prince Edward Island, it is the sole source. Over half the population of the US depends on groundwater for drinking water supplies; households in rural areas are especially dependent. Yet, in 1984, the Environmental Protection Agency knew of over 7,000 wells that had been closed or contaminated by groundwater pollution and 16,000 toxic waste dumps with the potential to contaminate groundwater.

Waste dumps pose threat to wells

Given that groundwater is such an important source of drinking water, it is surprising how little most of us know about it. Groundwater does not flow through the ground in streams nor is it stored in huge underground caverns. Instead, most groundwater fills the tiny spaces between soil particles or the cracks and pores in rock. Underground areas which hold substantial quantities of water in the soils or rocks are called "aquifers". These aquifers are the true sources of the cold, clear water from springs and wells that is held in such high esteem by those searching for a purer alternative to tap water.

Groundwater in aquifers can be of high quality

Groundwater has earned a reputation for purity. It is often safer than surface water because a natural filtration process purifies it as it passes through the layers of soil and rock. When humans introduce large amounts of sewage, fertilizers or toxic chemicals into the system this cleaning process can no longer keep up and the groundwater becomes contaminated.

Filtration can become ineffective

Pollution of groundwater

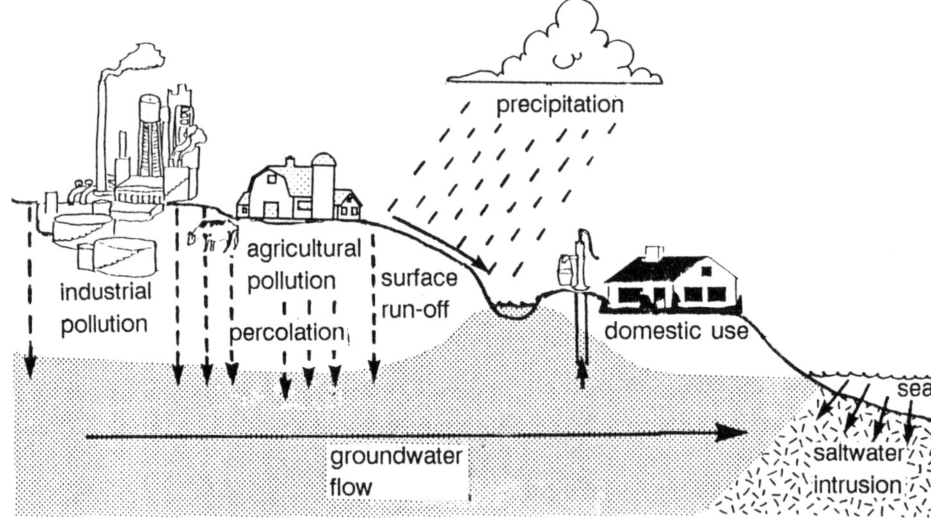

Pesticide residues in groundwater

Pollution of groundwater resulting from widespread use of pesticides and fertilizers, particularly nitrogen, is a major cause for concern. Widespread increases in nitrate levels and pesticide traces in drinking water supplies have been noted in numerous rural areas of North America and Europe. For instance, in 1984, 125 groundwater sources in England and Wales (supplying water for 1.8 million people) contained levels of nitrates which exceeded the UK's drinking water standards.

Traces of fumigant closes wells in US

An example of the entry of pesticides into groundwater is the case of DBCP (1,2-dibromo-3-chloropropane). This chemical was used extensively as a soil fumigant in California between 1950 and 1977. Its use was then banned because it was found to cause sterility in male workers in various production plants after exposure to low doses. As a result, chemical companies stopped production of DBCP and began recalling it from retailers and consumers. However, the problem did not end there. The contamination of groundwater by DBCP caused the closure of about 1,000 drinking water wells in the US.

Industry, home wastes pollute urban water

While agricultural fertilizers and pesticides can contaminate groundwater, they usually do so in areas of relatively low population. In contrast, groundwater pollution from industrial and domestic wastes or leaking underground chemical storage tanks tends to occur very near to where millions of people live. A survey of wells in California showed that the most polluted ones were in the Los Angeles area.

Chemicals from landfills enter groundwater

Millions of tonnes of industrial, building and power station wastes are disposed of in landfill sites every year. If chemicals in these wastes come in contact with water, some can dissolve and find their way into groundwater. There are of the order of 100,000 landfill sites in Canada and the US. There is no consistent monitoring of many of these waste disposal sites to determine whether or not they are contaminating groundwater and other water sources. This is particularly true for abandoned waste dumps. For example, there are over 19,000 abandoned hazardous waste sites known in the US, and some of them may be contaminating groundwater.

Chemical plant contaminates town's water

There is a long-standing groundwater contamination problem near Elmira Ontario which also illustrates the lasting nature of this problem. In 1989, the drinking water of the town was found to contain a suspected carcinogenic chemical, one of the nitrosamine family (see Chapter I). This chemical was then found in wells over 60 km away. Water from half the wells serving Elmira was declared unsafe, and water had to be trucked into the community. The source of the contamination was found to be a chemical plant which has operated in the town for over 50 years, and which is the town's largest employer. The contamination of groundwater centred under the plant will take decades to clean up.

Underground gas storage tanks common

Nearly every gas station has an underground storage tank. In the 1950s and 1960s, many thousands of tanks were buried across Canada, providing a convenient way of storing fuels or chemicals. Most of them are used for gasoline, diesel or fuel oil, but some are for storage of other types of chemicals. The majority of these tanks are thus buried at locations such as retail gas stations, bus lines or farms. Storing explosive or flammable materials like gasoline underground is much safer than storing them above ground.

Leaks from tanks contaminate water

The problem is that most of the tanks are made of steel. As time passes, they can rust and eventually leak. Up to 10% of the underground storage tanks now in use in Canada are leaking. This represents anywhere from 7,500 to 20,000 tanks - and the numbers are expected to rise dramatically as the average age of underground tanks in Canada increases. Because they are underground, it is very difficult to monitor these tanks for leaks. Besides, where leaks have been discovered, it has usually been too late to prevent damage - as little as one litre of gasoline can render one million litres of water unfit for drinking.

Government regulations for storage tanks

Governments in Canada are taking action to prevent the leaking of underground storage tanks. Ontario has a program of tank replacement, where tanks over 17 years old must be replaced with new tanks which have greater resistance to leakage. Because Prince Edward Island is very dependent on groundwater for drinking, it has had very strict regulations for underground tanks for years.

Aquifers stay polluted for a long time

Clearly, pollutants can, and do, leach through the soil into aquifers. Unfortunately, once polluted, aquifers stay that way for a long time. Groundwater migrates through the small pores in the rocks at rates of only several metres per year - once a contaminant has entered these pores, it can take years to flush out. Chemical degradation of pollutants is extremely slow in the cool dark conditions present in groundwater. In contrast to polluted surface water, which can be treated or aided in cleaning itself, contaminated groundwater is basically beyond our reach.

Prevention is the best solution

Even if all sources of pollution were removed immediately, it would take decades for contaminated groundwater to clean itself. Recognising the lasting nature of contaminated groundwater, many argue that working towards prevention of groundwater pollution is far more important than devising methods to cope with it.

OIL SPILLS

Oil spills disturb base of food chains

Fresh groundwater can become contaminated with petroleum products through leaking underground storage reservoirs. In the oceans and seas, oil spills are the most dramatic sources of contamination by petroleum products. The oil spill in the Persian Gulf during the war in Kuwait is the most serious and most recent illustration of this. It will be some time before the details and hence any reasonable estimate of the consequences of this spill will be known. Recent estimates put the size of the slick at two to three million barrels instead of the seven million originally feared. However, one thing is as certain in the Gulf spill as it is for every oil spill; although attention is usually focused on the obvious and immediate threats to birds and higher marine animals, the fate of the Persian Gulf ecosystem really depends on the spill's effects on the tiny creatures at the bottom of the food chain.

Some oil pollution is natural

There are many other ways oil can pollute water. Nature spills a fair amount of oil into waters itself. All oil and gas occurs naturally in the earth, but it is usually trapped deep beneath the crust, only occasionally bubbling up to the surface through fissures. Alberta's famous Athabasca tar sands are one example of a natural occurrence of oil that constantly pollutes a waterway. These tar sands contain approximately 40 billion tonnes of oil - as much as all other oil resources in North America. This oil, soaked into the sands of the area, is constantly oozing into the region of the Athabasca River surrounding Fort McMurray.

Dependence on oil threatens marine life

Nature's contribution to aquatic oil pollution pales beside that of modern societies. As we have grown increasingly dependent on oil in both our industrial and personal lives, we have greatly expanded our extraction and transportation of petroleum products. The pollution of waters, particularly marine waters, has been one side-effect of this activity. Pipelines and offshore drill rigs leak small quantities of oil into waters fairly regularly. Ships contribute a considerable amount of oil to the oceans when they wash out their ballast tanks. However, the spills caused by supertanker accidents are by far the most sensational examples of oil pollution in the world's oceans.

Oil spill incidences are inevitable

The effects of oil spills on marine ecosystems are not very well understood, but they seem to be quite localized and vary from spill to spill depending on the weather conditions, amount of oil spilt and whether the waters are open or enclosed. In the 45 years since World War II, the oil-consuming nations of the world have greatly expanded their use of ocean routes to transport crude oil. They have also, predictably, compiled a long list of accidental oil tanker spills. The most recent addition to the list is very familiar to North Americans. On 24 March, 1989 the supertanker Exxon Valdez ran aground on Bligh Reef near the coast of Alaska. Eight of the 11 cargo holds of the Valdez were ruptured in the accident, spilling 35,000 tonnes of crude into the cold, clear waters of Prince William Sound.

Harm to wildlife is obvious

As with most oil spills, the immediate and obvious effects were the deaths of thousands of sea birds and hundreds of sea otters when they were coated with the thick, gooey oil. The spill also covered hundreds of kilometres of shoreline with oil, sometimes over ten centimetres deep. In an oil spill, such effects usually last for a few months - after that most surface oil has either been cleaned up or washed out to sea.

Important effect on plankton not obvious

Unfortunately, the less visible (and less treatable) effects of an oil spill like the Valdez accident are more important to the fate of the ecosystem. Oil dispersed in the water column can kill microscopic plants and animals such as plankton that are the foundation of the aquatic food chain. Once in the water column, oil also threatens larval and juvenile fish and fish eggs - this could have severe economic consequences for the Alaskans who rely heavily on fishing for income and sustenance. In previous spills, oil has been found in the water column months after all surface oil is gone. Scientists fear that in the cold, arctic waters of Prince William Sound, this situation will persist longer than it does when spills occur in temperate waters.

Effects remain after many years

Finally, studies of past accidents show that when oil settles to the bottom, it can decimate the populations of bottom dwellers like shellfish and crabs. Unfortunately, evidence of continued stress on these aquatic inhabitants has been found six to ten years after a spill. All told, biologists and oceanographers estimate that the environmental effects of oil spills can continue 10 to 20 years after the event, and they do not know if affected ecosystems ever return to their pre-spill state. The delicate arctic ecosystem of Prince William Sound is no exception to this rule.

Clean-up procedures cause harm

The hot water often used to clean beaches contaminated by oil spills can cause environmental problems, perhaps even greater than the oil itself. The hot water used in Prince William Sound sterilized many parts of the beaches, killing mussels, barnacles and other forms of life. The resulting oily water gathered in tidal pools where it damaged sensitive organisms. From an environmental point of view, it may be better to allow nature to clean the oil residues off the rocks and sand in its own good time.

Direct harm to humans not well studied

The hazards of oil spills to human health are also not well studied. The risks of direct harm from the oil are probably only relevant to those involved in the clean-up. Crude oil is primarily a mixture of hydrocarbons, some of which may be toxic; benzene is one such toxic (carcinogenic) compound. Although the lighter of these compounds evaporate within a few days (the VOCs, see Chapter I), people exposed to the crude immediately following an accident could inhale some carcinogens or absorb them through exposed skin. Of course, this potential is known, so that most (if not all) workers wear protective clothing designed to guard against this.

Health risks possible via food chains

In the long term, it is unlikely that the oil itself represents any real danger to people. However, people eating contaminated fish could be at risk. In the case of the Valdez spill, the aboriginal Alaskans fit into this category - over 80% of their food is drawn from Prince William Sound. No one is really sure what the consequences to their health will be, since little research has been done on this phenomenon.

Dependence on oil is basic problem

In a broad context, oil spills are inevitable occurrences in today's oil-dependent technological society. Oil spills may differ in the specific events surrounding the accident, but they have the same fundamental causes: a worldwide dependence on fossil fuels and the untold benefits we derive from their use; and a reliance on oceans and seas as the principal routes for transportation. As long as these circumstances exist, so will oil spills.

CONCLUDING REMARKS

Three types of water pollution

Water pollution can be considered to be of three major types: pollution by micro-organisms such as viruses, bacteria, and other infectious organisms; pollution by chemicals which are toxic to humans or to other species; and another type of chemical pollution, by nutrients, which alter the ecology of bodies of water by accelerating the growth of aquatic plants and algae.

Biological pollution in Third World

Biological water pollution certainly presents the most risk to human health, especially in Third World countries. This serious hazard is however not addressed in any length in this book on chemical pollutants, even though Chapter V discusses the important role that chlorine has played in providing safe water for drinking, swimming and other uses.

Pollutants move between air-water-soil

Water pollution by toxic chemicals is discussed in the next three chapters: chlorine and its compounds, pesticides and metals. Chemical pollution of water is in many cases primarily due to deposition from the air, or to leaching of chemicals from the soil. Most of the toxic chemicals found in Lake Superior come from the air rather than direct inflows from streams and sewers. Groundwater is an important part of the freshwater cycle, and it is intimately related to soil pollution. Chemicals can enter groundwater from the soil, or be removed from the water through adsorption onto soil particles. Such problems can not be neatly categorized as being air, water or soil pollution.

Nutrient pollution major threat to water

A major part of this chapter discusses the third type of water pollution, that by nutrients such as phosphates and nitrates. Such pollution is not a serious immediate risk to human health, but it can wreak havoc with ecosystems in lakes and estuaries. Eutrophied bodies of water are also less useful for recreational and industrial uses, especially those uses involving fish or other forms of marine life. Steps have been taken to reverse eutrophication in several large lakes around the world. The problem of eutrophication in estuaries and other coastal regions may have been underestimated in past decades and may now have to be more seriously addressed. There is a limit to the vastness of the oceans as a sink for our sewage and we seem to be approaching that limit in several regions of the world.

Better sewage treatment a priority

Biological contamination may not be a serious health risk in developed countries such as Canada. One reason for this low risk is our choice not to use water that has been biologically polluted for recreation; beaches are simply closed when faecal coliform counts increase. There is no reason why, in a developed country, we can not clean up our own and our animals' sewage to the extent that most beaches in urban areas would be safe from biological pollution. This will require considerable resources in some areas where sewage treatment has had a low priority.

Chronic health effects difficult to assess

The risks to human health from low levels of toxic chemicals are difficult to assess. With an increase in the sensitivity of chemical analysis, we are finding more and more potentially hazardous chemicals. Some of these chemicals have probably been present in water for decades or even centuries, perhaps even in much greater concentrations. It is only now that advances in analytical chemistry make their detection possible.

Small adverse health effects add up

Ingestion of very small amounts of toxic chemicals over a very long time (chronic exposure) can result in negative health effects, but these may be too small to be detected in health data (see Chapter VIII). Even if the relative increase in health problems is very small, in a population of millions of people the overall health risk can add up to be considerable. Studies of the health effects of PCBs in the Great Lakes (see Chapter V) suggest that the regular ingestion of fish is the greatest exposure route, because fish have biomagnified the concentration of many chlorinated compounds and some metals (see Chapter VI). The risk is greatest to foetuses and infants so that women who are pregnant or who are breast feeding should limit their intake of fish.

Wildlife in danger if toxics persist

Because of biomagnification, species of carnivorous fish and fish-eating birds are at the greatest risk from pollution by persistent toxic chemicals in water. Herring gulls experienced a decrease in breeding success which was traced to high PCB levels. Very high mercury levels have been reported for older carnivorous fish in some lakes. This makes it inadvisable to eat such fish on a regular basis.

Risks from oil spills to ecosystem

The direct risk to human health from oil spills is small relative to the risks it presents to the ecosystem. Oil spills can have a catastrophic effect on some species of fish, birds and sea-mammals. It can take several years before some of these species re-establish themselves after a spill. The economic consequences can be considerable. They include lost earnings to the sea-food industries and the cost of cleaning up.

Recommended Reading

"Canadian Water Quality Guidelines", Canadian Council of Resource and Environment Ministers, 1987 (with subsequent updates).

"Fact Sheet 4: Water", Environment Canada. Supply and Services Canada, 1990.

"Nor Any Drop to Drink", B. Headlan. *Harrowsmith*, pp 41-49, March/April 1991.

"Water 2020", Science Council of Canada Report No. 40, 1988.

CHAPTER V
CHLORINE AND ITS COMPOUNDS

Chlorine is useful and hazardous

Many of us are familiar with chlorine as a household bleach, or as disinfectant added to swimming pools and drinking water. Some of us, including the 220,000 residents of Mississauga, Ontario, are familiar with chlorine in a different role: as a potentially deadly chemical. When a tank car carrying chlorine was punctured during a train accident in November of 1979, the threat posed by the leaking chlorine resulted in one of the largest evacuations in North America. For six days Mississauga was a ghost-town while workers struggled to control the situation. Fortunately, it ended without human injury.

Chloride ion is common and essential for life

The element chlorine is very common on the earth in its ionic form, the chloride ion. It is the most common ion present in ocean water. Chloride ion is a common and essential part of all living organisms. It is present in our bodies; our digestive juices are to a large extent hydrochloric acid. We add chloride ions to our food in the form of table salt.

Many different useful chlorine compounds

Chlorine in many other forms is also present in scores of products and processes encountered in daily life. It is a testament to human ingenuity that the potential of chlorine has been realized in such a wide variety of applications: food additives, paints, adhesives, plastics - the list goes on and on. Chlorine and its derivatives have brought enormous benefits to life in the industrialized world. But it seems that some of these benefits have been accompanied by unforeseen costs.

Many well-known chlorinated pollutants

Many of the chemicals now entrenched in the vocabulary of pollution - such as DDT, PCBs, CFCs, dioxins, 2,4-D, and trihalomethanes (THMs) - are chlorinated compounds. Sometimes these chemicals enter the environment directly via waste effluent, surface run-off or atmospheric emissions. In other instances, chlorinated pollutants are formed unintentionally as by-products. This is the case for the formation of THMs during water chlorination, and of dioxins during chlorine bleaching of pulp and improper combustion of PCBs.

CHLORINE GAS

Chlorine gas first used for bleaching

Chlorine gas was first discovered by a Swedish chemist in 1774. This pungent smelling, yellow-green gas was considered for some years as little more than a fascinating and dangerous by-product of certain combustion reactions. When, about 30 years later, chemists learned to produce chlorine more predictably, its usefulness as a bleaching agent was recognized and then increasingly utilized.

Gas damages lungs; used in warfare

By that time, the toxicity of chlorine was also becoming well-known. As any swimmer who has been in an over-chlorinated pool can attest to, it is very irritating to the skin, eyes and mucous membranes. If inhaled, chlorine is extremely harmful to the delicate linings of the respiratory tract and lungs, causing initial symptoms of tightness in the throat and constriction of the chest. Higher exposures lead to a horrible death; victims essentially drown as their body responds to the chlorine by filling their lungs with fluid. Chlorine gas is much heavier than air, so that it tends to form a cloud near the ground unless there is a wind to disperse it. Because of these properties, chlorine was the first gas used for chemical warfare in World War I.

Table salt is primary source of chlorine

Chlorine in nature is nearly always found combined with other chemicals. The most common chlorine compound on earth is sodium chloride (NaCl) - familiar to everyone as table salt. The abundance of sodium chloride helps explain why it is the most common source for extraction of the large quantities of chlorine used in industry. The most efficient and widely used method of obtaining chlorine from salt is called electrolysis, a process where an electric current is passed through a salt-water solution. The development of electrolysis near the end of the 19th century allowed large-scale manufacture of chlorine and started its importance as an industrial chemical.

Chlorine use pollutes in many ways

As with most elements that we extract from the earth, chlorine is innocuous in its common natural state. The manipulation of such elements creates useful chemicals, as well as pollutants, out of harmless precursors. In the case of chlorine, the production process itself has caused pollution in an unexpected way. In North America it was discovered in the 1970s that significant amounts of mercury were leaking from the electrolytic cells used for chlorine production and entering waterways through waste water effluent (see Chapter VII). As a result, the "mercury" cells have now largely been phased out.

Transportation of Chlorine

Transportation from centres of production

The location of chlorine production plants is largely determined by easy access to substantial rock salt deposits and a dependable supply of electricity. Chlorine production facilities must also be large to be economical; there are thus only about 13 such plants in Canada. As a result, much of the chlorine produced must be transported overland to reach its consumers. The necessity of transportation over long distances is by no means unique to chlorine production; it is the reality for most modern manufactured chemical products. Continual transportation of potentially dangerous substances, whether by air, land or sea, eventually leads to inadvertent spills of these chemicals. The risk of such accidents can be lowered, but never completely eliminated (see Chap VIII).

Transportation is potentially hazardous

Even though chlorine is a gas at room temperature and pressure, it is usually stored and transported as a liquid. It can be liquified either by cooling it to less than -34°C or by keeping it under a pressure of about ten atmospheres. Not only is liquid chlorine easier to transport, it is also easier to use than its gaseous form. The most common mode of transportation for chlorine is by rail in reinforced tank-cars under pressure. In the US, there are over 9,000 tank cars in service. Every year each car makes 12 to 16 trips, and every year, about 30 of them are involved in train wrecks. Few of these accidents lead to chlorine emissions, but when they do, there is a great potential for harm to people. If a pressurized chlorine car is punctured and heat is present, eg from a fire, the chlorine vaporizes, spreading out in a dense, toxic cloud that lingers near the ground, threatening the health of anyone who inhales it.

The Mississauga train derailment

For example, near midnight on Saturday, the 10th of November, 1979, a 106-car freight train derailed near Mississauga, Ontario. It was carrying tank cars of styrene, caustic soda, butane, propane and other solvents as well as chlorine. A tremendous explosion signalled that many of the flammables were ablaze. A punctured tanker filled with 90 tonnes of chlorine was lying upended on the track, surrounded by eight burning propane cars. The intense heat from the fire caused the steel in the chlorine car to burn spontaneously and about 70 tonnes of chlorine gas was released. Fortunately, the terrific heat and updraft caused by the blazing propane created a chimney effect that funnelled the chlorine gas up about 1,400 metres into the atmosphere. It was then dispersed and diluted by shifting winds, falling to earth in minute concentrations over a 100 kilometre radius. Widespread evacuation of the area was quickly ordered. By Sunday morning, virtually the whole of Mississauga was evacuated, as well as parts of the nearby town of Oakville.

No loss of life in Mississauga

Six days after it began, the event was over. Because of favourable weather conditions and a quick response by authorities, this accident ended with no loss of life or reported injuries. Ironically, it was probably the propane fire, the most dramatic aspect of the accident, that played the largest part in preventing this disaster from turning into a tragedy. By sending the chlorine high into the air, the fire effectively dispersed the gas over a wide area and gave people time to evacuate.

Chlorine spills have caused deaths

In other nations, however, chlorine tanker accidents have ended with more tragic results. In 1978, in Youngstown, Florida, leaking chlorine gas from a 90-tonne derailed tank car killed eight people and injured 138 others. Perhaps the worst spill in history occurred in August of 1981 in Mexico. A freight train with 32 tank cars of chlorine derailed about 100 metres from the train station in the small town of San Luis Potosi, population 400. Seven chlorine tank cars lost all of their contents and official reports listed thirteen people dead due to inhalation of chlorine vapour.

Spills force re-evaluation of regulations

This accident caused the Mexican government to change its regulations, no longer allowing such a concentration of chlorine cars in a single freight train. Similarly, the Mississauga derailment focused criticism and scrutiny upon Canadian transport regulations. The accident raised many questions such as why trains were permitted to carry combinations of both toxic and flammable materials and why there was no information about the order in which cars were coupled. No one knew a chlorine car was amongst the propane cars until they smelled its characteristic odour.

Assessment of transportation risks

Consequently, the Grange Commission was set up to consider the risks involved in the transportation of dangerous goods and assess the regulations for rail transport of such materials. In Canada, the regulation of transportation of hazardous materials is shared by the provincial and federal governments. International and interprovincial aspects are under federal jurisdiction, except in the case of road travel, where the task is delegated to provincial transport boards. All railways in Canada are governed by the federal Railway Act, except for intraprovincial railways in Ontario and British Columbia.

There will always be some risk

Transportation regulations have been reassessed and improved over the years, and they will undoubtedly continue to be revised. However, as long as we continue to transport hazardous chemicals like chlorine, there will always be some risk of damage to human health and the environment from accidental spills. In the end it is a question of balance, weighing the risks against the benefits and the costs of alternatives - and choosing accordingly.

Chlorine in Water

Wide use of chlorine for bleaching

Chlorine has been recognized as a powerful bleaching agent for a long time. By the end of the 19th century, the main commercial use of chlorine gas was in the manufacture of powder for bleaching paper and cotton textiles. This was to keep up with public demand for whiter products. Today, chlorine is still used extensively as a bleaching agent both in household and industrial use. For instance, in Canada, the single largest use of chlorine is as a bleaching agent by the pulp and paper industry. The formation of dioxins in bleaching with chlorine is discussed in a later section.

Chlorine kills algae, bacteria

Chlorine is an effective disinfectant; it functions as an algicide and a germicide. Its disinfecting power makes chlorine the predominant method for purifying water, both in swimming pools and for drinking. The chlorination process was first introduced in England in 1904. Today, about 99% of all municipal water supplies that are chemically treated use chlorine at some stage of the disinfection process. During this phase of water treatment, the water is made safe to drink by killing any harmful bacteria or viruses that may be present. Besides preventing algal growth, the chlorination of water also removes iron and manganese and controls taste and odour. Chlorine is all the more attractive because it is relatively inexpensive and easy to handle. It can be applied under controlled conditions and is measurable in the water supply at all points in a distribution system. Chlorination of water supplies has essentially eliminated the scourge of waterborne diseases such as cholera and typhoid in many communities.

Chlorine maintains disinfection

Equally important is the disinfectant's role in preventing any further contamination during the time that the water is travelling through the distribution system. Water in a large city can travel for up to five days before reaching the kitchen tap. Bacteria are quick to multiply, so that five days is plenty of time for any organisms that escaped death in the purification plant to proliferate. As well, any breaks or leaks in the water mains could introduce new organisms. A good disinfectant must thus be able to continue to protect water even after it leaves the treatment plant. Chlorine is unique among disinfectants for possessing this characteristic.

Chlorine's potency as a disinfectant arises because it undergoes reactions with water to form powerful oxidants that can pass into microorganisms and destroy them. This is not unlike the action of another familiar oxidant. Hydrogen peroxide (H_2O_2) oxidizes cells and bacteria as it fizzles in a wound and thus disinfects it. Chlorine reacts very readily with many substances; this quality makes chlorine valuable to industry, but it has also created some problems in water chlorination.

Chlorine forms oxidants that kill bacteria

The first indictment of water chlorination occurred in the 1970s when researchers discovered that the chlorine was reacting with substances in the water to form several potentially carcinogenic compounds. Chlorine can react with trace pollutants entering water through waste effluent from industrial activities. For instance, the manufacture of pulp and paper and certain herbicides discharge waste waters containing traces of phenols. If the phenols are still present during water treatment, chlorine reacts very quickly with them, producing various chlorinated phenols, and some of these cause cancer in rats (see Chapter VI). Fortunately, even the most minute traces of chlorophenols make the water taste so unpleasant that it is unlikely that such contaminated water would be ingested in any appreciable amount.

Reactivity of chlorine causes problems

Improved methods of water analysis resulted in the discovery of new hazards due to water chlorination. Attention is now focused on the unintentional formation of a set of chemicals called trihalomethanes, or THMs. During the chlorination process, chlorine can react with organic matter, such as decaying plants, to form THMs that were not present in the raw water sources. Some of the THMs, such as chloroform, are proven carcinogens in rodents and suspected carcinogens in humans (see Chapter VIII). Another new class of potentially hazardous compounds have recently been discovered in drinking water, the dihaloacetonitriles. The biological effects of these chemicals are now being studied. It is likely that even more hazardous chemicals will be detected in drinking water as analytical chemists refine their instrumentation.

Chlorination by-products can be hazards

Chlorine itself has never been shown to be carcinogenic, but legitimate concern has been raised by the potential health threat posed by exposure to carcinogenic chemicals formed during chlorination of drinking water. Some studies have shown slightly elevated cancer levels in communities that chlorinate their water compared to those that do not. In one case, epidemiologists (see Chapter VIII) prepared atlases showing the geographic distribution of cancer. This "geocancerology" indicated high cancer rates in parishes in Louisiana that drew water from the Mississippi River. An analysis of water supplies in this region found evidence of the presence of many chlorinated compounds.

Possible carcinogens in drinking water

In response to these results, both the US and Canadian government agencies initiated surveys to measure chloroform at various water treatment plants. The study conducted by Health and Welfare Canada attempted to correlate levels of this chemical with water source and treatment procedures in 70 municipalities. The concentrations of chloroform varied from below the detection limit to 0.12

Studies of level of chloroform

parts per billion (ppb). The amounts were almost always below the maximum acceptable limit of chloroform set by the federal and provincial governments for long-term consumption.

No proven health hazards of chloroform

The report concluded that: "on the basis of our present knowledge, it cannot be concluded that the consumption of water, containing the levels of halomethanes found in this present survey, represents a health hazard". Another fact that helps put the risk from exposure to ppb levels of chloroform in perspective is the following. Until the mid 1970s, many children's cough syrups contained several percent chloroform as a cough suppressant, a larger dose than would be received from water even after many years. Even so, this does not invalidate people's concerns over the health implications of even very low levels of THMs in drinking water. The long-term effects of exposure to low levels of chloroform on human health remain uncertain. Currently, the risks are estimated from animal studies (see Chapter VIII). Perhaps more importantly, little is known about the effects of life-time exposure to minute levels of combinations of chemicals; chloroform is just one of many trace chemicals found in water.

Chlorination saving many lives

At the same time, there is no question that chlorination of water has saved millions of lives over the past century in countries like Canada. Throughout history, water has been a confounding problem in environmental health as both a reservoir and transmitter of human disease. To this day, water treatment in many developing countries is inadequate or non-existent. In these countries, the death toll from waterborne diseases is enormous. The United Nations estimates that about 15 to 20 million babies and five million adults die every year worldwide from diseases such as typhoid fever, cholera and amoeboid dysentery. The incidence of waterborne diseases in most of the developed countries today is practically nil. Chlorine disinfection of water deserves most of the credit for this.

Alternative disinfection methods

There are alternatives to chlorine disinfection, including the use of ozone and ultraviolet radiation. These can be even more effective in disinfecting water than chlorine, but all of them are more expensive to use. Ozonation, in particular, is being used increasingly for water treatment in a number of regions, including France and Quebec. However, none of the alternative disinfectants, including ozone, possess the residual disinfecting quality of chlorine. Thus, even if an alternative is used, some form of chlorine has to be added to keep the water clean in the pipes while the water travels to the user.

Effects of alternatives unknown

The advantage of using alternatives is that only low doses of chlorine are needed for residual, post-treatment disinfection and most organic matter is out of the water so that there is less chance of forming THMs or other chlorinated compounds. On the other hand, the potentially harmful effects, both immediate and long-term, of these alternatives are not yet well understood. Substituting one set of problems for another seems to happen often in a technological society.

CHLORINATED COMPOUNDS

Chlorinated compounds very common

The chlorine used in water treatment accounts for only five percent of the total chlorine market in North America. The chemical industry consumes about half of all the chlorine produced for use in a wide array of products. Chlorinated compounds are commonly used as solvents in the chemical and electronics industries as they are particularly good at dissolving substances that do not dissolve in water. The pesticide and plastics industries are also large users of chlorine, accounting for about one third of the market. The plastic polyvinyl chloride (PVC) is commonly found in flooring, pipes, siding and many other applications. Many pesticides are chlorinated compounds; all three of the pesticides featured in Chapter VI (DDT, 2,4-D and PCP) are chlorinated compounds. Compounds containing chlorine are also used as anaesthetics, pharmaceuticals, detergents, household chemicals, dry cleaning fluids, rocket fuel, perfumes, etc.

Entry into the environment

Such widespread use of chlorinated compounds causes them to enter the environment in a variety of ways. For instance, cleaning solvents get into waterways through waste effluent (see Chapter IV). Other compounds are liberated through the natural degradation of products that contain chlorinated compounds. Still others are deliberately released into the environment. The story of the chlorofluorocarbons (CFCs) used as propellants in aerosols is discussed in detail in Chapter II.

Compounds are persistent pollutants

Many chlorinated compounds are very persistent pollutants. Soil organisms cannot break down many of these compounds, and they are quite stable in cool, dark regions such as aquifers (see Chapter IV). In addition, they are often lipophilic, ie they prefer to move from water to body fats. This means many of these compounds can pass up food chains, increasing their concentrations in the higher levels of the food chain many times. This is why recognizing a danger and ending all production of a given compound does not necessarily solve the problem. The longevity of many of these compounds ensures that past contamination will be responsible for their environmental presence for many years to come. A detailed discussion of the story of one group of such persistent environmental pollutants follows.

PCBs (Polychlorinated Biphenyls)

PCBs a vilified class of compounds

Polychlorinated biphenyls, or PCBs, just might be the most vilified class of chemicals of the past twenty years. The unique qualities that these chemicals possess made them valued tools for many applications that provided social gain - until harmful consequences of their use were discovered. One lesson that can be learned from the PCB story, is that PCBs typify the risks of embracing new technologies while lacking the ability to foresee a full range of consequences.

PCB toxicity greater with more chlorines

Although PCB molecules can become quite complex, the basic components of PCBs are simple: hydrogen, carbon, and chlorine. PCBs were first synthesized in 1881, but were not applied industrially until 1923. While the number of carbon atoms is constant, the numbers of hydrogen and chlorine atoms vary. As a rule, the toxicity of a PCB increases as the chlorine content increases. PCBs are almost insoluble in water, are unaffected by acids, bases or corrosive chemicals, have high boiling points, and can be heated to quite high temperatures without decomposing. Chemical stability and low electrical conductivity are the two properties that have made PCBs very useful for many industrial purposes, particularly in the electrical industry. PCBs are excellent non-conducting fluids for capacitors and transformers; in fact, nearly all of the capacitors and transformers built between 1930 and 1980 contained PCBs.

Many uses for PCBs worldwide

The electrical industry was the principal user of PCBs in Canada, but these chemicals served a variety of other functions as well. PCBs were used in many countries as heat-transfer and hydraulic fluids; as flame-retardants; in lubricating oils; as plasticizers in synthetic resins, rubbers, paints, waxes and asphalts; and they were used in carbonless copy paper. Oil containing PCBs was even sprayed on roads to suppress dust.

PCB stability an environmental problem

Thirty years after the introduction of PCBs, people realized that the qualities that made PCBs so ideal for so many industrial uses had an unexpected and unwanted corollary. In 1966, PCBs were detected in the feathers of eagles in Sweden. It seemed that not only were PCBs persistent in the environment, but they also biomagnified in food chains. This path could take them through freshwater and marine plants, through birds, fish and other animals, leading eventually to humans. For the first time, people recognized that the same stability that made PCBs so popular also made them serious environmental contaminants. Although a few plants and bacteria have been found which can degrade them, PCBs are even more resistant to biological and chemical breakdown than DDT (see Chapter VI).

Many routes from product to pollutant

PCBs may enter the environment from a number of sources. Leaking transformers and capacitors, burning of carbonless copy paper, landfill leachate, or leaking tankers are all possible sources of PCBs. Release of PCBs into the environment was deliberate when they were used as dust suppressors on roads, or as agents to extend the active life of insecticides. Accidents during transport and electrical transformer fires are also potential routes for PCBs to infiltrate the environment. PCBs have occasionally caused concern in indoor air pollution; video display terminals and defective fluorescent light ballasts might be origins of such pollution. A study found that air in commercial, industrial and residential buildings contains levels of PCBs at least one order of magnitude higher than outdoor air.

PCB levels higher in urban areas

Since 1966, traces of PCBs have been found throughout the world in soil, snow, the atmosphere, milk, food, plankton, and sediments in rivers and lakes. It seems that PCB levels are higher near industrialized and urban areas. A map of PCB concentration in the environment mimics a map of population density. PCBs are lipophilic; thus they bioaccumulate in animal or human fatty tissue. Traces of PCBs have even been found in mothers' milk in regions as seemingly remote as the Canadian Arctic.

Health Effects

Adverse health effects at high exposures

As with so many pollutants, the consequences of the presence of PCBs in the environment are still being debated. Their effects at higher concentrations are more conclusive. Studies using birds and monkeys showed that PCBs reduce reproductive success, liver function and affect immune and enzyme systems. Some accidental exposures to PCBs or compounds formed from PCBs have had tragic effects for people.

PCB poisons rice oil in Japan, Taiwan

The first accident occurred in Japan in 1968, on the island of Kyushu. Cases of headaches, vomiting, fever, eye discharges, visual disturbances and a severe skin rash called chloracne were reported on an epidemic scale. One thousand and fifty-seven patients were treated for these symptoms by 1971. The epidemic was traced to contaminated rice oil containing 2,000 to 3,000 ppm of PCBs. Investigation revealed that a leak in the heating system of a rice oil producer in Kitakyushu City had contaminated the edible oil with PCBs. A second mass poisoning occurred in Taiwan in 1979 from oil contaminated with PCBs. The 300 people who were affected showed toxic symptoms similar to those observed in Japan.

Higher cancer rates after accident

The Kyushu accident did more than focus public attention on the hazards of PCBs; it also raised the fear that PCBs cause cancer. Eleven years after the event at Kyushu, 51 victims had died, and of the 31 for whom a cause of death was established, 11 were from cancer. Such a cancer rate is above average, and it seemed to link PCBs with liver cancer.

PCB link to cancer not proven

However, while PCBs are known carcinogens in rats, no studies have yet found a direct relationship between exposure to PCBs and human cancer. A 1982 Health and Welfare Canada study concluded that skin disorders, abnormal liver function, growth depression in children and respiratory impairment are associated with exposure to high levels of PCBs. Although many people believe that a zero level of PCBs is the only safe level, others point to evidence that shows no harmful health consequences from exposure to low levels of PCBs. Three studies of workers who were occupationally exposed to PCBs for over 40 years did not find any indication of ill effects. These studies, conducted by the US National Institute for Occupational Safety and Health (NIOSH), examined workers who had higher than normal PCB levels in their blood. They found that the cancer rates among the workers were slightly lower than the national average.

More research raises doubts of hazard level

The issue of PCBs' toxicity to humans is further complicated by studies made in 1981 of the oil from the Kyushu accident. Testing of the oil from the accident found that it contained traces of the highly toxic chemicals polychlorinated dibenzofurans (PCDFs). The PCDFs are formed when PCB molecules incorporate oxygen atoms. Similar concentrations of these furans were related to toxic effects seen amongst European workers in 1970. This evidence suggests now that PCBs were not directly responsible for the illness and death following the Kyushu accident. They may be indirectly responsible, as tests found that used PCBs from a transformer contained furans in concentrations close to the levels found in the oil from the Kyushu accident. Laboratory tests then showed that PCBs can degrade to furans when heated to temperatures of around 300°C in the presence of oxygen. Conditions like these exist in transformer "hot spots".

Greater indirect hazard of PCBs

In a practical sense, it seems irrelevant whether PCBs themselves or their by-products are responsible for the toxic effects observed in studies and in accidents. If known toxins such as furans are formed under certain circumstances from use of PCBs, then the dangers posed by these chemicals must be considered part of the risk of the use of PCBs.

Health effects for infants near Great Lakes

The consumption by pregnant women of fish with high levels of PCBs seems to be the greatest risk to human health from PCBs in the environment. The foetus is very susceptible to chemical insults, and PCBs pass through the placenta to the foetus. A study in 1984 showed that infants born to mothers who consumed fish from the Great Lakes contaminated by PCBs had lower birth weight, smaller head circumference and shorter gestation. A follow-up study gave results which suggested some deficiency in the performance of seven month old infants in tests of visual recognition memory.

Evidence remains inconclusive

As discussed in Chapter VIII, there are many problems in obtaining clear evidence from epidemiology for negative health consequences of exposure to a chemical. A correlation between a lag in infant development and fetal exposure to PCBs exists, but many other variables that could not be controlled could affect the outcome: these include demographic background, maternal smoking, consumption of alcohol, etc. It is in any case prudent to try to minimize the exposure of infants, and especially foetuses to PCBs.

Control of PCBs

PCB production reduced and ended

By the early 1970s, the presence of PCBs in the North American environment had been solidly confirmed and pressure was mounting for government and industry to take action. In 1972, the only manufacturer of PCBs in North America took a proactive stance by voluntarily cutting their PCB production in half, from 40 to 20 million kilograms per year, and restricting the sales. This restriction affected those uses that were most likely to result in losses to the environment, such as hydraulic fluids, heat transfer fluids and plasticizers. In 1976, plans were announced to cease the production of PCBs by the fall of 1977.

PCB uses restricted in US and Canada

The Canadian government was motivated to act in 1973, when the Organization for Economic Cooperation and Development (OECD) recommended that member countries restrict the use of PCBs to closed systems such as transformers and capacitors. PCBs were then the first substances to be regulated under the Environmental Contaminants Act passed in 1976. Initially, their use was restricted to new electrical equipment, then to the servicing of existing electrical equipment. New use of PCBs was finally banned on July 1, 1980.

Concentrations in humans declining

Control measures such as these have brought reductions in the numbers of people having high levels of PCBs in their bodies, at least in the US. In 1972, 2.7% of the population had levels of PCBs greater than three ppm in their body fat. By 1977, this figure had increased to eight percent. However, it had decreased to one percent by 1981.

Release of PCBs continues

Nevertheless, the PCB problem continues. Of the 40 million kilograms of PCBs imported into Canada before their manufacture was ended in 1977, about 24 million kg can be accounted for. There are still many transformers filled with PCBs, and about six million kg of concentrated PCB liquids lie in storage awaiting disposal. As for the other 16 million kg of PCBs, it must be assumed that they are either present in low concentrations in mineral oils or they are already in the environment. Unfortunately, PCBs will continue to be released into the environment for some time. Of the PCBs recorded as in use or in storage, it is estimated that 18 kg are lost every day.

PCB spill on Ontario road sparks concern

By 1984, most of the regulations for handling PCBs were in place, and the public concern with PCBs had decreased. Then two accidents happened during the eighties which ensured that PCBs would remain prominent in the consciousness of Canadians. The first was a spill of PCB-contaminated oil over a 100 kilometre stretch of the Trans-Canada highway near Kenora, Ontario. Four hundred litres of transformer oil containing 56% PCBs were spilled along the highway, requiring a great deal of time and money to clean up. It seems a sign of the times that the quantity of PCB-oil involved in this accident was very small compared to the amount deliberately used twenty years earlier to keep the dust down on roads.

Fire of oil with PCBs leads to evacuation

It was, however, the fire in St. Basile-Le-Grand, Quebec on August 23, 1988 that riveted national attention on PCBs. The dramatic fire of oil containing PCBs in a storage depot in that town made international newspaper headlines for weeks. The accident, marked by confusion and fear, resulted in the evacuation of 3,500 people. For three weeks, these people were kept from their homes while the government checked the entire area for traces of toxic chemicals, particularly dioxins and furans. The temperatures experienced in the fire at St. Basile-Le-Grand could have caused the formation of these compounds, and their possible presence in the toxic cloud that was formed during the blaze was the real concern of government officials. By November, almost all of the 5,600 people involved in the fire were given a clean bill of health. About 100 people showed abnormal liver enzyme levels, but these were not believed to be due to PCBs or the related dioxins or furans. Monitoring continued on 450 people. About half of these were firefighters who had fought the blaze without masks or protective clothing.

Accident accelerates PCB phase-out

The after-effects of the fire were felt for many months. Media criticism of national and provincial PCB regulations brought attention to the fact that, for years, Quebec had not had any regulations for PCB storage. It seems that even after safety rules were enacted, they were not enforced. The Quebec Environment Minister admitted that he had known for years that the warehouse in St. Basile-Le-Grand was unsafe but nothing had been done. Reportedly, the unguarded warehouse did not even have a danger sign to warn people of its contents. Indeed, many of the firefighters said they did not know what was in the storehouse and thus had fought the fire without protection against chemical contamination for days. This accident was a major factor in accelerating the federal government's phase-out program for PCBs.

PCB users to find substitutes by 1993

Until that time Canada, like many countries, had adopted a natural attrition policy for the phasing out of PCBs for authorized uses. This meant that the use of a PCB-containing product could continue until the end of that equipment's life. This was supposed to take until about the year 2000. However, a surprisingly short time after the St. Basile-Le-Grand fire, the federal environment minister announced that Canada would adopt an accelerated phase-out program. Under the new program even existing authorized users of PCBs will have to find substitutes by 1993.

Public fears over disposal of PCB waste

The St. Basile fire also outlined a common problem encountered when trying to deal with any substance considered to be hazardous waste. The PCBs in this fire were in storage because they were awaiting disposal. Unfortunately, the very concerns that people feel about PCBs make them cautious about allowing any kind of disposal facility to be built in or near their communities. Public trust of the government in matters such as disposal of toxic wastes seems to have eroded over the past decades. Many people find it difficult to believe officials when they give assurances that a particular waste treatment is safe. Yet leaving PCBs in storage is not a solution to the problem. Clearly there are risks involved in storage as well as in disposal.

Incineration is highly efficient

Although PCBs are difficult chemicals to destroy, there are several methods that seem to work well. The most effective method of destroying concentrated PCB wastes is high temperature incineration. The stability of PCBs means that combustion temperatures must reach at least 1,000°C before these chemicals are completely broken down. In early 1981, the US Environmental Protection Agency (EPA) approved the construction of two incinerators for PCB destruction. In Europe, ocean incineration has been carried out successfully for some years. At a site in the Gulf of Mexico, a special incineration ship disposed of PCB waste with an efficiency of 99.99%. However, in recent years, protests against ocean incineration have increased because of concerns about ocean pollution and the dangers of accidental spills during transport of wastes to ports.

Public wary of incinerators

Norway uses cement kilns to destroy PCBs. In Ontario, the Ministry of the Environment conducted plant trials in a modified cement kiln. The trials were held in Mississauga despite municipal opposition and achieved 99.986% destruction. But public opposition has been so strong that a permanent site for an incinerator in Ontario has not been found. To date the only operating incinerator in Canada for PCBs is located in Swan Hills, Alberta. It was licensed in 1988 and is working to capacity, destroying 12,500 tonnes of PCB wastes annually. It will take several years for this incinerator to destroy all the PCB wastes stored in Alberta. Only then will it accept wastes from other parts of Canada.

The search for more acceptable PCB disposal facilities has resulted in the development of mobile incinerators. Using incinerators that go where the problems are will greatly reduce the chances of accidental transportation spills. Equally important, mobile incinerators relieve some of the anxiety associated with having a disposal facility permanently located in any one community. The first mobile unit is already in place in Goose Bay, Labrador.

Mobile incinerators used for PCBs

DIOXINS AND FURANS

Dioxin has been called the "toxic chemical of the eighties". The discovery of traces of this family of chlorinated chemicals throughout the environment has made it a source of widespread public concern. When the word "dioxin" comes up as a pollution issue, people often assume that the substance involved is 2,3,7,8-TCDD (tetrachlorodibenzo-para-dioxin), the most hazardous of all the dioxins. But this is not always the case. In reality, dioxins are a family of 75 related compounds known as polychlorinated dibenzo-p-dioxins (PCDDs). Furans, closely related to dioxins, are part of a family of 135 compounds called polychlorinated dibenzofurans (PCDFs). Furans and dioxins are similar in their chemical structure, with furans having one less oxygen atom. Their toxic effects are also similar, so that they are usually grouped together for discussion.

"Dioxin" refers to a class of chemicals

The accidents at St. Basile-Le-Grand and Kyushu, Japan, make it clear that dioxins and furans are a part of the PCB problem. Eliminating PCBs would certainly ease the environmental pollution from these chemicals. Unfortunately, improper PCB combustion is just one source of furans and dioxins. Members of this family of chemicals are formed as by-products of many different human activities. At one time, TCDD and the other dioxins were considered to be contaminants that were formed unintentionally during several chemical manufacturing processes.

Many sources of dioxins and furans

A study released in 1978 suggested that dioxins were also formed in trace amounts during combustion, including forest fires. By 1981, other research studies had found dioxins in incinerator fly ash, fireplace soot, motor vehicle exhaust systems and cigarette smoke. Their production essentially depends on the type of material burned and the conditions of combustion. The traces of dioxins are usually stuck onto small particles, which carry them into the environment. Thus, emission controls are important for large incinerators and power plants. The emission of dioxins from such sources varies widely, with concentrations ranging from 0.001 to 1,100 ppb.

Dioxins also formed in fires

Lately, the dioxins entering waterways through waste effluent from pulp and paper mills have been featured in media coverage. Small amounts of dioxins are formed in mills that use chlorine to bleach their pulp. Of more immediate concern to many, dioxins are also present in minute concentrations in many bleached paper products to which consumers are exposed daily.

Dioxins from paper mills

Dioxins in humans from eating beef

Recent studies suggest that about half of the dioxin found in humans comes from eating beef. Dairy products and fish account for about 15% each. The major route by which dioxins enter our food chain seems to be by these compounds sticking onto particles, settling on grass or on water, and being consumed by animals. In the case of fish, the concentration is magnified by the food chain.

Dioxin release in Seveso accident

Dioxins are also formed as by-products in the manufacture of certain pesticides. One such source of dioxins received worldwide attention in 1976. TCDD is a by-product of the manufacture of trichlorophenol, an intermediate chemical used to produce the herbicide 2,4,5-T (one of the ingredients of Agent Orange). On July 10, 1976 there was an accident at the ICMESA plant in Seveso, Italy. On this day, a safety disc in the reactor ruptured, releasing a cloud of chemicals containing trichlorophenol and an estimated 0.5 to five kg of the contaminant TCDD. As a result of the accident, animals died and many people experienced nausea, headache, and skin irritation. Eventually, the town was evacuated and work began to clean up the hundreds of acres of land that were contaminated.

No death of people from dioxin

The evacuation was in response to legitimate concerns over the health effects of TCDD. This compound is 500 times more poisonous than the better known poison strychnine and 10,000 times more poisonous than cyanide, at least for guinea pigs (see Chapter VIII). Nevertheless, no human deaths have as yet been attributed to dioxin poisoning either from low-level environmental exposure or from industrial accidents, including Seveso.

Health Effects of TCDD

Health effects in animals from dioxin

TCDD residues accumulate in liver and fat cells. Clinical symptoms associated with exposure to the chemical include: atrophy of the kidney, degenerative changes of the liver, ulceration of the stomach, haemorrhaging of the gastro-intestinal tract and other organs, and atrophy of the thymus and lymphoid organs and tissues. Occupational exposure can lead to symptoms of systematic poisoning: damage to the liver, heart, kidney, spleen, central nervous system, and pancreas; memory and concentration disturbances; emphysema; and depression. The skin disease chloracne is caused by the body trying to rid itself of the poison through the skin. It is extremely contagious and can pass rapidly through a worker's family, simply from contact with contaminated clothing. The metabolization of dioxins has been found to be slow and represents a small fraction of that ingested. Most of the dioxin is excreted quickly, or stored in fat and then excreted gradually.

Possible health effects from Vietnam

TCDD is a proven teratogen in animals, ie it causes birth defects in the offspring of those exposed to it. In Vietnam, birth defects such as cleft palates and kidney weaknesses were observed, but it can not be stated conclusively that dioxin-contamination from the herbicide Agent Orange was responsible. TCDD is also listed as a suspected carcinogen. The substance causes malignant tumours in rats, but there is still disagreement about the carcinogenicity of dioxin to humans. American Vietnam veterans are claiming damages for injuries they say were caused by the dioxin in Agent Orange.

Occupational exposures to TCDD

The most extensive human epidemiological data come from workers in trichlorophenol and 2,4,5-T plants who have been subjected to TCDD exposure, usually as a result of releases due to excessively high reaction temperatures. This occurred in the US (1949, 1964), West Germany (1953), Holland (1963), and Britain (1968). The most common complaint was chloracne. A recent German study involving workers from a pesticide plant showed that those with more than 20 years of service had twice the normal death rate from cancer. The plant closed in 1984 because it produced dioxin-rich residues. Dioxin was however not the only chemical which could have been responsible.

Recent data suggest risk overestimated

A recent study by the NIOSH suggests that the health risk to humans exposed to dioxins may have been overestimated. Workers from the chemical industry who had first been exposed to dioxin at least 20 years earlier were considered in groups. The group with blood levels of dioxins 90 times higher than those in the general public had no detectable ill effects. A second group with 500 times the general blood level had a risk 50% greater than normal of dying from cancer.

Long-term studies of Seveso effects

The Seveso accident in 1976 provides another opportunity to study the toxic effects of dioxins on humans. It is known that the less developed the organism, the more susceptible it is to toxic substances. Because of this, the 250 women exposed to dioxin who were in the first trimester of pregnancy had abortions. Thirty of these aborted foetuses were examined for defects; 29 were found to be normal, the 30th was too badly damaged to tell. The children of Seveso are being watched especially carefully for signs of ill health. A mass screening of 42,000 children of the area has revealed about 600 suspected and 134 confirmed cases of chloracne. Some trends in health effects do seem to be emerging, notably an increase in spontaneous abortions and impairment of the peripheral nervous system, but the epidemiological data from monitoring Seveso residents is incomplete and inconclusive. A substantial number of animals died or had to be killed to prevent the consumption of contaminated meat, but no human fatalities seem to have been caused by the dioxin release. Another decade of monitoring is likely to be needed to get more definitive results.

Cleaning dioxin from the environment

The ecological effects of the accident are yet to be determined. In Britain and the US, as well as in Italy, scientists have been engaged in decontamination research. Various methods have been suggested: treatment with an emulsion of olive oil which may facilitate the alteration of dioxin by sunlight to a less harmful compound; incineration of contaminated soil; and bacterial degradation. Some contaminated soil from Seveso which was shipped away ended its journey in a French slaughterhouse, showing up weaknesses in toxic waste regulations in Europe. Dioxin is disappearing from the topsoil near Seveso, possibly due to degradation caused by sunlight. The dioxin found below the surface has an estimated half-life of ten years.

TCDD in the Environment

Dioxin in chemical disposal sites

Some chemicals, in particular trichlorophenol (TCP), can contain dioxins as impurities. Regulations on the dioxin content of commercial chemicals mean that the contamination must be reduced or the TCP must be discarded. There has thus been a requirement for disposal of dioxin-contaminated waste. The Love Canal in the US was one such disposal site; it contained 200 tonnes of waste. When the clay barrier on the disposal site was broken during the building of a school, the area was contaminated by various toxic chemicals and was eventually evacuated. The Hyde Park site in Niagara Falls, New York, contains 3,300 tonnes of TCP and is believed to be the largest known store of the chemical.

Leakage from dumps into the Great Lakes

Hyde Park and other sites are known to be leaking dioxins into the Great Lakes system, Lake Ontario in particular. These leaking waste repositories in the Niagara Falls region are considered responsible for the TCDD detected in herring gull eggs on the Canadian side of the Great Lakes at levels ranging from 44 to 64 parts per trillion (ppt). Monitoring of fish in Lake Ontario found TCDD levels in sports fish at between 20 to 25 ppt; in commercial fish, eels and smelt, levels were less than 20 ppt.

Recommended TCDD levels very low

One part per trillion is a very small amount; it is equal to one drop of liquid in about 10 large swimming pools. Permissible levels of TCDD in fish have been set by Health and Welfare Canada at 20 ppt. This is based on an annual consumption of 13 pounds of fish and provides a safety factor of 2,000 from first measurable effects of teratogenicity and carcinogenicity. The US Food and Drug Administration has developed a guideline of 50 ppt for sport fish; the New York guideline is 10 ppt for six ounces per week consumption; Ontario's is 20 ppt for four ounces weekly. Toxicity studies have shown fish to be the most sensitive organisms to dioxins and furans.

Threats to drinking water

The long-term leakage of TCDD from waste disposal sites in the Niagara Falls, New York region poses a potential threat to the five million people who draw drinking water from Lake Ontario. The clean-up plans for the landfill sites near the Niagara River will only slow down the rate of chemical seepage. TCDD has yet to be detected in North American drinking water. The detection limit for this chemical in water has now been lowered to 0.005 ppt. Traces of TCDD as high as 0.03 ppt have been found for raw water from western Lake Ontario, but water treatment processes seem to reduce the concentration to below detection limits in tap water.

CONCLUDING REMARKS

Love-hate relationship with chlorine

Industrial society has had a "love-hate" relationship with synthetic chemicals containing chlorine. Industrial chemicals such as PCBs seemed to be ideally suited for some uses: they were chemically nonreactive and thermally stable. These same properties are now the reason for the great concern about PCBs in the environment; since they are nonreactive they accumulate. Because of this, PCBs were classified as first priority hazards to the environment and received a great deal of media attention.

PCBs make easy headlines

A tragic occurrence such as the Kyushu oil incidence, and perhaps also a good headline label ("PCBs"), make PCBs in the environment and PCB destruction a major media issue. The news media often showed a skull and cross-bones symbol when discussing PCB issues after the Kenora and the St. Basile accidents in the 1980s. Yet there is no evidence that exposure to PCBs has caused any substantial ill effects, let alone death, in Canada. There is however some evidence from the US that exposure of foetuses to PCBs through their mothers' ingestion of fish contaminated with PCBs has had detrimental effects on their development.

Real impact of PCBs, dioxins still unknown

The evidence that PCBs and dioxins have harmed species such as fish and falcons is substantial. It is thus important to reduce the release of these substances into the environment to the lowest possible levels. Zero levels may eventually be approached for PCBs as there may be no natural sources for these chemicals. Dioxins, however, are produced in nature through fires, so that we will never reach zero levels. With increasing sensitivity of analytical equipment, we will find more and more contamination by dioxins - not because there is more contamination, but because we are able to detect smaller and smaller amounts of contamination.

Exaggerating risk causes anguish

The danger of grossly exaggerating the risks of chemicals is illustrated by the Seveso incident. In that case, "the pen was mightier than the chemical". The anguish caused by unfounded fears of ill effects from exposure to dioxin seems, in retrospect, to have caused more harm than any chemical effects. Most of the abortions carried out were probably wanted babies. The manager of the plant at the time of the accident was also later assassinated by the Red Brigade in Italy after being "tried" and found guilty of the dioxin contamination.

Little evidence of harm from chloroform

Anxiety about chlorinated compounds such as chloroform in drinking water seems to be on the increase. Yet even though chloroform was used as an anaesthetic and an additive in toothpaste and cough syrups for many years, there seems to be no evidence of harm to humans. The benefits of chlorination of water seem to far outweigh the risks of the formation of chlorinated compounds in that process.

Small risks can add up over time

It is nevertheless prudent to try to minimize the amount of compounds such as chloroform in the environment. They have been found to be carcinogenic in animals, and it is unwise to let them build up. Even if the risk of cancer to an individual human from exposure to a small amount of chloroform is very small, when many people are exposed for many years, the overall health effect on a population may be considerable.

How safe is "safe enough"?

The section about the transport of chlorine illustrates the problems encountered in transportation of hazardous materials generally. The degree of precautions to be taken illustrates the problem of the "planner's dilemma". We can insist on stricter and stricter safety standards: better railway tracks, more railcar inspections, etc. Most such regulations will increase the cost of transport, and these costs will be passed on to the consumers. How safe is "safe enough"?

Recommended Reading

"Dioxin: Research Needs for Risk Assessment", C.C. Travis and H.A. Hattemer-Frey. *Chemosphere*, pp 729-742, vol 20, nos 7-9, 1990.

"Contaminants in Herring Gull Eggs from the Great Lakes", SOE Fact Sheet 90-2, Environment Canada, 1990.

"The PCB Story", Canadian Council of Resource and Environment Ministries, 1988.

"A Review of the Toxicology and Human Health Aspects of PCBs (1978-1982)", Health and Welfare Canada, 85-EHO-113, 1985.

"Polychlorinated Biphenyls (PCBs) - Fate and Effects in the Canadian Environment", Environment Canada, EPS4/HA/2, 1988.

CHAPTER VI
PESTICIDES

"There was a strange stillness. The birds, for example - where had they gone? . . . No witchcraft, no enemy action had silenced the rebirth of new life in this stricken world. The people had done it themselves."

A quote from "Silent Spring"

Thirty years have passed since Rachel Carson wrote this passage, trying to alert a society that was showering the land with pesticides. Her book "Silent Spring" presented a scenario of the environment that could be expected if pesticide use were to grow unabated. It inspired people to consider how pesticides affected the total environment. "Silent Spring" was a very major factor in the formation of environmental lobby groups and of new government institutions, in the introduction of new regulations for pesticide use, and in the development of a new awareness of the hazards of chemicals in the environment.

The start of an environmental movement

Many aspects of pesticide use have changed in the last three decades, but we continue to rely on them to improve agriculture, forestry and public health. We have introduced a diverse arsenal of pesticides for killing various forms of life (the suffix "-cide" comes from the Latin word "cida", to kill); insecticides to kill insects, herbicides to kill weeds, rodenticides, fungicides, and other -cides. These pesticides combat organisms which may cause illness or which compete with humans for such things as food, forest products or even roses in the backyard. Many millions of people have been spared the ravages of malaria because of the use of insecticides to control disease-carrying mosquitoes. The food supply of the world is as plentiful as it is, partly because of the pesticides used to protect crops. This much is clear. What is not clear is how residues of these chemicals in the environment affect us over a longer time period.

Pesticides have many beneficial effects

Since all living organisms have some points of similarity, it should be assumed that all pesticides can be dangerous to all living species. Thus although fungicides are meant to be toxic to fungi, they are likely to have some toxic effects on insects and humans as well. Some pesticides have a very "broad spectrum" of action, while others are quite species-specific. Modern pesticides are usually designed for a specific target organism and are planned to have little effect on other forms of life.

Broad-spectrum or species-specific

This chapter begins with a general overview of pesticides, covering such topics as history, types of pesticides, pollution of the environment, health effects, and regulations. One of the most famous of all pesticides, the insecticide DDT, is used in conjunction with this overview to help illustrate the problems of pesticides. The remaining sections discuss in more detail the herbicide, 2,4-D and PCP which is used primarily as a fungicide. These chemicals are involved in controversies in Canada. They illustrate the problems involved in regulating pesticide use in such a manner that we derive the benefits from the use while keeping the risks to human health and the environment to an acceptable level.

Overview followed by 2,4-D, PCP

HISTORY OF PESTICIDES

Pesticides have been used for a long time

Humans have used pesticides for millennia. Early civilizations in what is now the Middle East used sulphur compounds to control insects and mites as far back as 2500 BC, and the Chinese have used fumigants since 1500 BC. However, a pest-control industry only started to develop during the mid-eighteenth century. The French concocted a mixture of lime and sulphur to protect their grape vines from mildew in 1851, and in the US arsenic compounds were widely used to combat the Colorado potato beetle beginning in 1867. World War I marked a rapid increase in the development of chemical pesticides. The Russians used chloropicrin for gas warfare in 1916. Chloropicrin was also found to be a good insect fumigant, so that the remaining stocks were turned to this less violent use after the war.

DDT starts the modern era of pesticides

The modern pesticide industry really developed during World War II when DDT (dichlorodiphenyltrichloroethane) was widely used. Although DDT was first synthesized in 1874, its toxic effects on insects were not recognised until 1939. DDT is a powerful and persistent insecticide which is lethal to a broad spectrum of insects but only moderately toxic to mammals. It has a "lethal dose 50%" (LD_{50}) of 87 milligrams per kilogram of body weight (mg/kg) for mammals (rats), making it about as toxic as the nicotine found in cigarettes. Chapter VIII explains the concept of "lethal dose 50%". In 1943, DDT was used on a large scale for the first time in Naples, Italy, to prevent an outbreak of typhus in the allied army. By 1945, over 20 million kg of DDT had been produced in the US, largely for use in the war effort to kill flies, lice, mosquitoes and ticks. DDT was even incorporated into uniforms and used in powder form in soldiers' underwear.

DDT was used widely in the 1950s and 60s

As a result of wartime activity, several US plants had the capacity to produce the chemical at low cost. The stories of DDT's incredible wartime success in controlling malaria, typhus and other insect-borne diseases were well known to the general public. The result was overwhelming public support for this new pesticide and it soon became widely used for public health protection, household insect control, government spraying programs and agricultural use. During the 1950s and 1960s, DDT was used extensively in tropical countries as part of a malaria eradication campaign sponsored by the World Health Organization (WHO) to reduce the population of mosquitoes that spread the disease. WHO estimates that between 1950 and 1975, one billion cases of malaria were prevented and five million lives were saved through this campaign.

Popularity of DDT decreases

But in the 1960s, the public's affection for DDT had faded in the industrialized world. Evidence was growing indicating that the use of DDT was resulting in several unexpected and undesirable side effects. For instance, scientists discovered that some flies and mosquitoes had become resistant to DDT.

Insects develop immunity to pesticides

In natural populations, a small number of insects may be resistant to a given pesticide and are not killed by it. They in turn produce a high proportion of resistant off-spring, which are able to survive subsequent insecticide applications. The short lifespan of insects means that, with continuous use of a pesticide, large-scale resistance can develop within only a few years. This phenomenon, first observed with DDT, is a problem encountered with many pesticides currently in use. For chemical pest control to remain effective, it is often necessary to switch to new pesticides every few years to overcome the resistance problem.

DDT is toxic to many non-target species

DDT also illustrated the problems that can arise when using broad-spectrum pesticides. People recognized that the non-selective nature of DDT's action made it toxic to non-target animals such as frogs and fish. Furthermore, it affected many beneficial insects such as bees. It also acted against insect predators, without which pest populations were more likely to reach extreme numbers. One example of this was observed when DDT was sprayed to kill coddling moths. Apparently, the DDT was also killing predatory lady beetles, and the result was a population explosion of red mites. This type of situation set up a cycle where each outbreak of pests called for the use of more insecticide which, in turn, killed more predators. Thus, pesticide use had to spiral upwards just to maintain agricultural productivity.

DDT persists in the environment

Other properties of DDT were discovered which raised alarm about its potential environmental and human health effects. DDT, like the other "organochlorines" is a synthetic chemical, and there is not necessarily an effective natural process for breaking such chemicals down. They are thus very persistent and can build up and spread in the environment.

DDT bio-accumulates in the food chain

This realization was made more significant when it was discovered that DDT could "bioaccumulate" in the food chain because it, like many organochlorines, is stored in fatty tissue. When an animal with DDT in its body fat is eaten by a predator or scavenger, the pesticide dose moves along into the body fat of the new host. Predators thus accumulate higher and higher quantities of the pesticide. The levels of DDT were becoming quite significant in species such as predatory birds and fish at the top of the food chain. By the mid-1950s, DDT had also been found in the body fat of the general population.

ppm means parts per million, ppb per billion, and ppt per trillion

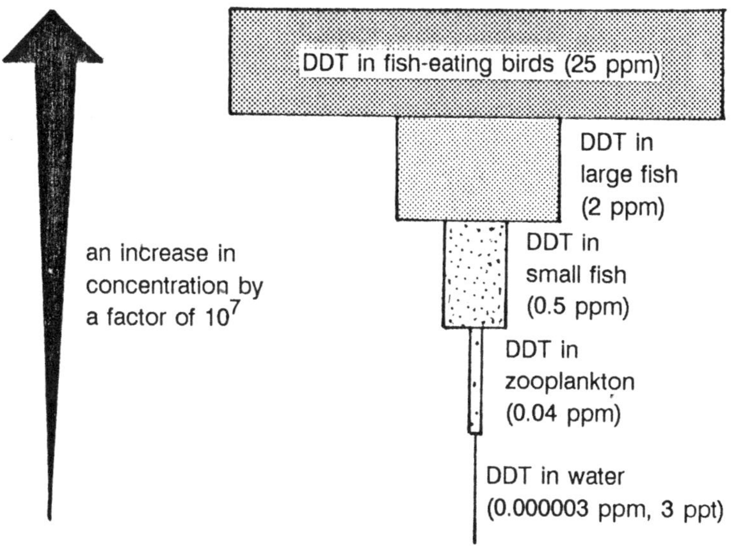

Bioaccumulation in a Food Chain

Concern about DDT leads to bans

Awareness of DDT's undesirable effects occurred at a time when both scientific and public concern was becoming focused on the environmental impacts of pesticides. This was largely due to the enormous popularity of Rachel Carson's book "Silent Spring", which was published in 1962. The result was a steady decline in DDT use by industrialized nations. In 1969, Sweden banned the use of DDT and by 1973, there was essentially a total ban on DDT in the US.

DDT is still used in tropical countries

DDT is now used mainly in tropical regions of the world to control insects carrying diseases or agriculturally harmful insects. It is rarely used for both purposes in the same area since agricultural use can enhance the rise of insect resistance which undermines the more important public health efforts.

Global use of pesticides has increased

Even though DDT has been phased out in the US and many other countries, the worldwide use of other pesticides has increased. Some two million tonnes (about half a kg per person) of pesticide products are now introduced into the world environment each year. The industrialized countries still use most of this, about 80%. In these nations, up to one-half of the pesticides used are for non-food crops, such as cotton and tobacco. Uses in Third World countries account for the remaining 20% of global pesticide consumption. About 70% of these pesticides used are for crops intended for export. The rate of pesticide use in the Third World has increased rapidly in the last decade as these countries have attempted to increase their agricultural productivity and health standards. Growth rates of pesticide use as great as 35% per year in Asia and 22% per year in South America have been noted.

TYPES OF PESTICIDES

Metal-based pesticides were once common

Before 1940, the most commonly used pesticides were inorganic, being compounds of arsenic, copper, mercury or other metals. Inorganic pesticides, such as arsenic insecticides or mercurial fungicides, are still used to a limited extent. Once released to the environment, the toxic metals in these pesticides will never degrade, although in time they may migrate to places where they cannot cause harm.

Hazards of inorganic pesticides

Serious consequences have resulted from the use of inorganic pesticides. For instance, mass poisonings caused by the consumption of grain seed treated with mercurial fungicides have occurred in Iraq, Pakistan and Guatemala. Mercury residues flushed into streams from agricultural uses and factories contribute to a high level of mercury found in fish in some lakes, although in other lakes, a very high level of mercury exists naturally. The topic of mercury and other forms of metal pollution is covered more extensively in Chapter VII.

Three types of organic pesticides

In the previous section, the discussion centred around DDT. It is an "organic" compound, which means that carbon is a major component, as in all biological compounds. Some naturally derived organic pesticides have been used for decades; for example nicotine extracted from tobacco, and pyrethrum derived from chrysanthemums. The newer synthetic organic pesticides fall into three main categories: organochlorines, organophosphates and carbamates. The different groups vary in their environmental persistence, toxicity and cost of production, but their mode of action is similar in that they all interfere with important enzyme functions.

Organochlorines build up in fatty tissue

Organochlorine pesticides, which include DDT, are usually persistent, do not dissolve well in water and often have a broad spectrum of insecticidal activity. These qualities can be advantageous, but they can also be environmentally hazardous. When ingested in small doses, they accumulate in fatty tissues in the body where they apparently cause no ill effect as long as they remain in the fat. However, when there are food shortages or when people go on a diet, these pesticides can re-enter the metabolic system in concentrated doses as body fat is consumed. As a consequence of these and other potentially harmful effects, many countries have regulations to control and restrict the use of organochlorine pesticides.

Organophosphates are less persistent

The majority of pesticides currently in use belong to the second group, the organophosphates. These pesticides show a diversity of biological activity and persistence, but in general they are less persistent and more toxic than organochlorines. For instance, the insecticide malathion is only slightly toxic with an LD_{50} of 1375 mg/kg, while parathion (also an organophosphorus insecticide) is extremely toxic (LD_{50} = 13 mg/kg). Organophosphates are easily absorbed through the skin and are toxic if inhaled, absorbed or swallowed. They are water soluble and are excreted in urine so that mild exposures at infrequent intervals do not accumulate in body tissue the way organochlorines do.

Carbamates are least persistent

Carbamates resemble the organophosphates in their biological action, as they also inhibit the working of enzymes in the central nervous system. In general, carbamates are the least persistent type of pesticide and they are rapidly metabolized and excreted. The insecticide aldicarb is extremely toxic (LD_{50} = 0.93 mg/kg), while some carbamate fungicides have LD_{50} values over 5,000 mg/kg. There are also many pesticides in common use which do not belong to the above four main groupings.

POLLUTION OF THE ENVIRONMENT

Aerial spraying can disperse pesticides

When pesticides are used for agriculture, forestry or public health, they are often applied by aerial spraying. When this method is used, only a fraction of the pesticide application may actually reach the target pest. This is not only because the spray is widely dispersed, but also because fine particles of the pesticide formulation can remain suspended in the air and come down in rain even a month afterwards. Using this method, it is inevitable that unintended areas will receive small doses of the pesticide. Similarly, it is quite likely that some people will come into contact with the spray. Aerial spraying is also subject to the random influences of air currents. Short and long range transport by prevailing winds means that pesticides can be carried tens to thousands of kilometres away from the target areas.

Agent Orange caused great controversy

Several cases of large-scale pesticide spraying operations have resulted in controversies regarding their health effects on the public. Perhaps the most notorious case is that of Agent Orange in Vietnam. Agent Orange was a combination of the herbicides 2,4-D and 2,4,5-T, and was quite heavily contaminated (0.1 to 47 ppm) with dioxin (see Chapter V). Large amounts of it were sprayed from 1961 to 1970 in an attempt to defoliate the jungle along the trails used by the Viet Cong to supply their forces in the south of Vietnam. During the spraying operation, US officials took no steps to protect either the people of Vietnam or the soldiers from the herbicides, even though 2,4,5-T was known to be a teratogen (meaning it could cause birth defects). Many Vietnamese civilians suffered toxic effects from acute exposure to Agent Orange, and Vietnam veterans claim that their health was damaged as a result of this campaign.

Worldwide concern about 2,4,5-T

In Canada, there have been many cases where pesticide use by the forestry industry has led to controversy. One example occurred in 1983 when a group of landowners in Nova Scotia charged a forestry company for subjecting them to a health hazard by spraying a 2,4,5-T/2,4-D mixture on the company owned softwood plantations near their properties. The landowners lost the court case, but the kinds of concerns they expressed were mirrored by people worldwide. As a result, many countries have either banned or severely restricted the use of 2,4,5-T. In 1983 the only North American manufacturer of this product announced that production of the herbicide would be discontinued. Today little or no 2,4,5-T products are used in North America. The only allowable uses in the US are for application to rice fields and rangelands where 2,4,5-T can be used to kill weeds that are toxic and teratogenic to cattle.

Controversy surrounding fenitrothion

Another case involved the aerial spraying of fenitrothion over forests in New Brunswick to control spruce budworms. In the early 1970s, a group of scientists published research that implicated this organophosphate insecticide as a factor in an illness known as Reye's Syndrome, which predominantly affects children. However, no definite link between fenitrothion and Reye's Syndrome has been established, and subsequent research seems to point to the over-the-counter drug aspirin as a more likely cause of Reye's Syndrome.

Restriction of pesticide types

Controversies such as these have resulted in bans on the spraying of some pesticides, especially the organochlorines. Such restriction on the use of insecticides can make it very difficult and costly to control some insect plagues. For instance, the worst plague of locusts in 30 years started in North Africa in 1987. Previously, the organochlorine insecticide dieldrin had been successfully used to control locusts without any reported human casualties. But because of environmental concerns about the persistence of organochlorines, the Food and Agriculture Organization (FAO) switched to fenitrothion which has a half-life of only three days in the environment. Massive quantities were used to combat the locusts in 1987-88, but it was not very effective in controlling the plague.

Pesticides contaminate groundwater

Contamination of groundwater with pesticides has been reported throughout the world. Drinking water in Ontario was found to be contaminated with the herbicide alachlor. Since alachlor has been shown to be an animal carcinogen, the risks from the contamination were considered high enough that the use of alachlor was suspended in 1985.

Chemical spills have led to gross contamination of water. A fire at a chemical warehouse in Switzerland in 1986 led to the escape of about 20 tonnes of pesticides into the Rhine River. The mass killings of fish were believed to be mainly due to the presence of atrazine in this spill. Atrazine is not biodegradable by microbes in natural waters; thus it persists for a long time.

Accidents can lead to severe pollution

Any river surrounded by intense agricultural development, such as the Rhine, is continuously subjected to pesticide pollution due to the run-off from pesticide-treated fields. Persistent pesticides entering groundwater and rivers eventually end up in lakes and oceans. In Japan, fish and shellfish were still contaminated with dieldrin in 1984, even though all of its uses were banned there in 1981.

Water pollution can persist for years

Soil can also be contaminated for a long time by pesticides. The answer to the mystery of how melons in California became contaminated with aldicarb may be soil contamination. About 280 people were made ill in the US by aldicarb contamination. This was surprising, since aldicarb is not registered for use on watermelons. It may have been illegally applied, but there is evidence that it could have entered the watermelons from aldicarb-contaminated soil. Four fields with contaminated melons had been used some years earlier to grow cotton, and aldicarb was used on the cotton crops. Some aldicarb residues may have been present in the soil years later, although, according to the manufacturer, it breaks down in the soil in six months.

Pesticides can persist in soil for long times

HEALTH EFFECTS

According to WHO, approximately one million people worldwide are poisoned each year by pesticides, 20,000 fatally. In industrialized countries, few fatalities arise from pesticide use. As an example, in the late 1980s, the yearly death toll from poisoning in Canada was about 400. An average of only six of these 400 were involved with pesticides, with only one of the six an accidental death rather than a case of suicide.

Few deaths in Canada due to pesticides

About half of the injuries and two-thirds of all pesticide related deaths occur in Third World countries, even though only 20% of the world's pesticides are used in these countries. If people in Third World countries were as aware of the hazards of pesticide use and adopted the same degree of safety measures as people in industrialized countries, the number of total deaths due to pesticides per year would drop by up to 60%. Many of the poisonings and deaths are from occupational exposure of farm and orchard workers, but a substantial number are due to deliberate ingestion of pesticides in suicide attempts.

Pesticide-related deaths in Third World

Death and injury statistics express primarily the result of acute (single large dose) poisoning. Epidemiological studies have been carried out which implicate occupational exposures to pesticides in increased death rates. For instance, a

Adverse health effects may be underestimated

study conducted in a district of the Philippines showed a 27% increase in the death rate among working rural men during a period where there was widespread use of insecticides by farmers in the region. The death rate for urban men not exposed to insecticides fell during the same period. Results like these suggest that the risks of pesticide use may be underestimated. Pesticide poisonings may be misdiagnosed because of the vagueness of early symptoms of acute poisoning which can resemble other ailments such as asthma, heat exhaustion, or gastroenteritis. Deaths attributed to such causes as stroke and leukaemia may actually be related to pesticide exposure to a significant degree - especially in the Third World.

Chronic effects of pesticides uncertain

When considering the issue of environmental pesticides pollution, the more subtle hazards of chronic or long term exposure to low doses of pesticides are probably more relevant to the vast majority of people. But the chronic effects of pesticides are difficult to determine. One reason is that, because the symptoms of chronic pesticide poisoning are very vague, they can perhaps never be separated from overall health statistics. Another problem is that measurement of very small amounts of pesticides in human tissue and food is difficult and expensive. Some decades ago it was impossible to measure part per billion (ppb) levels, as the sophisticated analytical instrumentation needed was not yet developed. It is much more difficult to detect trace amounts of such chemicals than to detect trace amounts of radioactive contamination.

Accumulation of DDT in human tissue

The case of DDT illustrates some of the difficulties encountered in trying to determine chronic effects of pesticides. As already mentioned, DDT is a fat soluble, water insoluble pesticide, so that it accumulates in the fatty tissue of people. During the 1960s and 1970s, DDT was commonly present in the environment. Through continual ingestion of small doses, the concentration of DDT and its metabolites in the fatty tissues of people throughout the world built up. Very high average levels were found for some groups: 55 parts per million (ppm) in India in 1965, 45 ppm in Texas in 1970, 40 ppm in Mexico in 1975. In the 1970s, the average level of organochlorine pesticides in human tissue in the US decreased markedly. The concentration of DDT and its metabolites decreased from eight ppm in 1970 to four ppm in 1979.

Little evidence of harm to humans from DDT

DDT and other organochlorine pesticides have been shown to produce liver enlargements and cancers in mice and rats when administered at the maximum tolerable dose (MTD) (see Chapter VIII for explanation of MTD). However, even during the height of DDT contamination in 1965, doses in humans were estimated to be well below the minimum effective dose at which tumours formed in mice. No epidemiological evidence has ever been found linking DDT with liver cancer in humans. In fact, there has been a continual decline in liver cancer rates in humans since DDT came into use in 1940, with no evidence of an increase in the 1960s or 1970s when the effects of DDT would have been expected to show up. There is also no clear epidemiological evidence of reduced fertility or foetal malformation in humans caused by DDT. There is, however, some evidence that occupational exposure to very high levels of DDT may have caused some temporary neurological changes.

Many of the other pesticides, such as those in the organophosphate group and the carbamates, are active only for a short time and do not accumulate in human tissues. But chronic, low level exposure which people would experience, for instance, from using household insecticides, could result in adverse effects to the nervous system. There is some evidence from Britain that many "summer colds" or "stomach bugs" and other ailments and allergies may be due to exposure to low levels of organophosphate pesticides.

Illness from organophosphate exposure

The risk of cancer to the US consumer from eating pesticide-contaminated food was recently assessed by the National Academy of Sciences (NAS). Using carcinogenicity estimates obtained from animal data for the 28 most common pesticides in use, they calculated that about 20,000 cases of cancer per year could occur in the US if foods containing the maximum tolerable level of pesticides allowed by the Environmental Protection Agency (EPA) were consumed. This rate of cancer formation would be about the same as that for radon gas (see Chapter III). The above estimate is a "worst case" scenario, where the pesticide residues are all at maximum allowable levels; in reality this would be very unlikely. However there are other risks which were not included, such as birth and genetic effects or neurological and immune system damage. The estimates also leave out the risks from pesticides other than the 28 that the NAS considered, and the risks posed by inert components in pesticides.

Study of cancer risk from pesticides

DDT residues are believed to have had serious effects on some bird and animal species. Reproductive success among the carnivorous species at the top of the food chain, such as gulls and falcons, has been especially badly affected. The residues bio-concentrate in these animals through the food chain. There is a possibility that these effects were in part due to polychlorinated biphenyls (PCBs), organochlorine compounds which contaminated the environment (see Chapter V). Before 1970, PCBs may have been mistaken for DDT by the analytical methods used in those days. There is however now good evidence that DDE, a breakdown product of DDT, is the primary pollutant responsible for eggshell thinning in many species of birds.

Birds suffer from exposure to DDT

REGULATING PESTICIDES

Pesticides have provided undeniable benefits to both public health and agricultural productivity. The effectiveness of DDT in diminishing malaria has already been mentioned. The high gains realized in agricultural productivity in the past three decades have also, to some degree, been due to the use of pesticides. In the past decade, India has increased its agricultural productivity so much that it now has a surplus of food. This surplus averted major famines when crops were unusually poor in 1987 due to droughts and floods.

Benefits of pesticide use

In industrialized countries, modern pesticide-assisted agricultural technology has resulted in large increases in yield. Western Europe and North America have surpluses of most agricultural commodities. Because of the high yields, less land is required for agriculture and more is available for other uses such as

Higher agricultural yield frees land

nature reserves, parks and urban and suburban development. If the US were to use the pre-pesticide agricultural technology of 1940 to produce the amount of food obtained in 1980, an additional area about the size of the land east of the Mississippi would have to be brought under cultivation. Alternatively, about 60% of the forests or 70% of the pasture land in the US would have to be converted to cropland.

Modern farming is dependent on pesticides

The effect on agriculture of a dramatic decrease in pesticide use is open to conjecture. One worst case scenario predicts that withdrawal of pesticides from agricultural use might result in the loss of two-thirds of all crops. To a large extent, industrialized countries have adopted agricultural practices which include a dependency on pesticides. In North America and Europe, farmers have actively pursued large-scale cultivation of fewer and fewer strains of high-yield crops. While these strains provide enormous gain in the short-term, their genetic uniformity makes them vulnerable to massive destruction by pests. In a sense, we are trapped into relying on pesticides to guarantee the safety of these monocultures.

Possibility of decreasing pesticide use

However, many people feel that this reliance on pesticides is short-sighted. Some studies suggest that pesticide-intensive agriculture, although giving high yields in the short-term, is not sustainable in the long-term because of problems such as the destruction of pest predators and the development of resistant pest species. The route to reduced pesticide use is not simply to stop using them; it must also involve changes in agricultural practices. Some scientists suggest that pesticide use could be reduced by as much as 90% without adverse effects on agricultural yields or the cost of food.

Regulations differ between countries

Each government judges the benefits and risks of pesticide use when drawing up regulations. Benefits and costs depend on the geography and the socio-economic condition of countries. Therefore the regulations for pesticides differ significantly from country to country. Because of this variation, a pesticide manufactured in one country may be banned there, but may be allowed in another country. In order to prevent dumping of such hazardous products, the FAO has recently adopted a resolution that pesticides be traded only with Prior Informed Consent (PIC).

Prior Informed Consent system introduced

Under the requirements of PIC, the exporting country must inform the importing country of the product concerned and its hazards. Only after the importer has received this information and given consent can the transaction be completed. A system such as PIC could have prevented the tragedy caused by the insecticide phosvel which was produced in the 1970s in the US. It was sold to Egypt, Indonesia and South Vietnam while in the process of being tested for use in the US, prior to approval for its use. In Egypt alone this insecticide was implicated in the death of one field worker and the poisoning of 65 others, as well as the deaths of several hundred water buffaloes. Production of Phosvel was later banned in the US when a large number of workers at the plant were found to have neurological disorders.

The Pesticide Action Network (PAN), a worldwide coalition of over 200 non-government organizations formed in 1982, has not only been campaigning for PIC, but also for the banning of 12 pesticides called the "dirty dozen". They believe that the risks associated with using these pesticides outweigh any benefits. One of the alternatives they suggest for the dirty dozen is to use Integrated Pest Management (IPM). This involves the use of a combination of diverse tactics: biological control, or the use of natural pest predators or diseases; culture techniques such as crop rotation; development of pest-resistant varieties; the rational and limited use of selective pesticides, etc. It would be possible to have a wholesome and reliable food supply with more sustainable agriculture using IPM rather than only "chemical" management of pests.

PAN lobbies for integrated pest management

2,4-D (2,4-DICHLOROPHENOXYACETIC ACID)

The herbicide, 2,4-D, is the most popular pesticide in Canada, ridding many lawns of dandelions and other broadleaf weeds, thus relieving the symptoms of many allergy sufferers. It has been used in agriculture for 45 years and became infamous as one of the ingredients in Agent Orange, used as a defoliant during the Vietnam War. This weedkiller is often the source of public controversy because many people fear that 2,4-D and its contaminants cause cancer and birth defects.

2,4-D has been used for decades

The chemical works by imitating plant growth hormones called auxins. When it is applied, 2,4-D is absorbed through the leaves and carried throughout the plant. There it disturbs the balance between growth and nutrient handling. The effect of 2,4-D depends on the dose applied: in low doses, the herbicide can act as a growth regulator; at higher doses, it stimulates explosive growth which exhausts and thus kills the plant.

Plants "grow to death"

The major uses of 2,4-D in Canada are for cereal farming, where it boosts yields by 10-15% by eliminating competing weeds; in forestry, where it is used to suppress the growth of less desirable hardwoods; and for weed control in lawns. Since 2,4-D is a hormone-like herbicide, it must be used with care near sensitive broadleaf crops such as tomatoes, grapes and tobacco. Compared to some newer herbicides, 2,4-D is not a very powerful pesticide. Several hundreds of grams per hectare are required to achieve the desired results whereas newer herbicides only require five to ten grams.

Use in cereal farming and forestry

Since 2,4-D and other phenoxy herbicides are broken down by heat, water, sunlight and microorganisms, they are not very persistent in the environment. The degradation takes two to four weeks and is affected by the rate of application, temperature, moisture conditions and the organic matter content of the soil. One of the breakdown products, 2,4-dichlorophenol, may produce dioxins in the presence of sunlight under alkaline (high pH) conditions.

2,4-D persists only for weeks in environment

Some dioxin found in 2,4-D

In October of 1980, Agriculture Canada scientists presented a paper at a conference in Rome, announcing the discovery of chlorinated dioxins in 2,4-D. There are several different methods of preparing 2,4-D. Each method produces characteristic impurities, with some producing dioxins. It should be noted that the most toxic of all the dioxins, TCDD, was not found in any type of 2,4-D. Agriculture Canada's response to the presence of dioxins in 2,4-D was to take action to see that all 2,4-D used from 1982 onwards was "dioxin-free". But the increasing sensitivity of analytical methods means it is now possible to detect extremely small amounts of such chemicals; it is thus not realistic to expect that any chemical is present in zero concentration. This realization led to an amendment of the regulation in August of 1981 to allow a maximum content of 10 ppb of any specific dioxin.

Steps taken to improve 2,4-D use

By 1990, progress had been made in upgrading the quality of 2,4-D being sold in Canada and in obtaining more experimental data, which supported the herbicide. Improvements were also made in the use pattern for 2,4-D. This was accomplished by eliminating certain volatile forms of the product to reduce environmental contamination and limit drift damage, eliminating uses that were unnecessary or inadequately supported by data, and defining more precisely appropriate dosage ranges.

Old pesticides regulated by old standards

Part of the reason that the hazards of 2,4-D were not investigated earlier is that it is an old pesticide. In the 1940s, safety assessments were mainly concerned with acute toxicity. Since 2,4-D has a low acute toxicity, it satisfied these safety requirements. Safety criteria have greatly expanded in the last 20 years to include: chronic (long term) toxicity, teratogenicity, carcinogenicity, as well as other aspects. As safety standards increased, new chemicals on the market were properly tested, but older chemicals such as 2,4-D were often neglected. Consequently, there were significant gaps in the scientific knowledge about its health effects. Not until the 1980s were efforts made by governments to examine 2,4-D in terms of hazards such as birth defects and environmental impacts.

No significant health risk from 2,4-D

A special committee set up by the Ontario Pesticides Advisory Committee of the Ontario Ministry of the Environment reviewed the human health effects of 2,4-D. They stated in a 1986 report that "the existing animal and human data are insufficient to support the finding that 2,4-D is a carcinogen and consequently finds insufficient evidence to conclude that existing uses of 2,4-D pose a significant human health risk." Agriculture Canada agreed with this review but encouraged users to apply the chemical with care and respect. Occupational exposures are much greater than exposures to the general public.

Aquatic plants, wildlife harmed by 2,4-D

As mentioned earlier, based on current scientific knowledge, 2,4-D is not a persistent chemical in the environment and it does not bioaccumulate. But some effects may occur in wildlife when habitats are sprayed with the larger dosages recommended for uses in forestry. Some aquatic plants are susceptible to the herbicide in low doses, and many wetland birds depend on these plants for survival. The problem is of concern especially in the Prairies where wetlands are now in short supply. Although 2,4-D is relatively non-toxic to bees, the destruction of bee forage by the herbicide could have an indirect impact on these insects.

Implications of ban on 2,4-D

The long-term future of 2,4-D in North America will depend on the results of new toxicity studies which regulators expect to receive in the next few years. The financial implications of discontinuing its use are considerable; Canadian sales are worth more than 450 million dollars a year, and world sales are in the billions. A ban on the use of 2,4-D for grain production would mean an additional cost of 130-200 million dollars per year for Canada from lowered agricultural productivity. Homeowners would be unable to purchase any broad-leaf pesticide over the counter because there are no replacements. The available alternatives for homeowners are non-chemical means of weed control such as pulling the weeds out by hand, or to allow the weeds to grow.

PCP (PENTACHLOROPHENOL)

Accidental contamination of food by PCP

Pentachlorophenol (PCP) is one of a group of synthetic chemicals called chlorophenols which has been used for wood preservation since 1936. It is itself toxic, and it can also contain various toxic dioxins. There have been several well-publicized cases of PCP inadvertently entering food. In 1977, the state of Michigan banned PCP after the chemical was found in cows which had either licked PCP-treated wood or had breathed the fumes. In that same year in Canada, a shipment of feed grains for consumption by livestock became contaminated with PCP because of improper cleaning of a the railway boxcar which had previously carried PCP. As a result, the livestock and milk were withheld from the market. Then in 1980, on the heels of a discovery of dioxin in herring gull eggs from the Canadian Great Lakes area, other dioxins were found in chickens in Guelph, Ontario and traced to PCP-treated wood.

Shift to kiln-dried lumber

Most PCP in Canada is used for wood preservation. Freshly cut softwood lumber is usually treated with PCP to protect the wood from surface mould and fungal deterioration during transportation and long storage periods. Currently, about 1200 tonnes of this imported pesticide are used yearly at pressure treating facilities designed for impregnation of industrial wood products. PCP is virtually the only chemical used in Canada for preserving the wooden poles used for power and telephone lines. In the wood protection industry, there is now a trend to use kiln drying instead of chlorophenols. The reasons for the shift are that wood with less moisture content is lighter and thus cheaper to ship, and that new US regulations require kiln-dried lumber for wood house construction.

Uses of chlorophenols

PCP has been used as a bactericide, slimicide, fungicide and a herbicide. It has served as an insecticide in paints, textiles, photographic solutions, hides and leathers, drilling muds, industrial cooling systems and pulp and paper mill systems. PCP is also used in health care and veterinary products and in disinfectants.

Chlorophenols disrupt energy flow in bodies

There are two sources of adverse health effects associated with PCP use. The first source is the PCP itself. Chlorophenols are toxic in that they interfere in the mode by which the body utilizes energy. In general, the toxicity of chlorophenols (CPs) increases with an increase in the number of chlorine atoms attached to the phenol. Since PCP is the most chlorinated CP, it is expected to be the most toxic. The lethal dose of PCP is similar for a wide range of organisms including mammals and fish, in the order of 100-200 mg/kg.

Dioxins, furans formed as impurities

The second source of health hazards associated with PCPs arises from some of the impurities formed during its manufacture. During the latter stages of PCP production, elevated temperatures are needed, and this favours the formation of chlorinated dioxins and furans (see Chapter V).

PCP is not a carcinogen

PCP has not been shown to cause cancer. A two-year study of the effect of PCP on rats did not find PCP to be carcinogenic when administered orally for long time periods at doses sufficiently high to cause mild symptoms of toxicity. At high levels pure PCP has been found to produce chronic effects, such as decreased body weight. There is evidence that PCP is toxic to rat embryos and foetuses but there is no indication that PCP is teratogenic. As a result, in 1987, Agriculture Canada proposed the following hazard warning for labels on all products containing PCP and other related chlorophenols: "...exposure to this Pest Control Product during pregnancy should therefore be avoided. Women who are or may be pregnant should not handle or apply penta or tetra chlorophenol. Employers should inform their workers of the risk associated with exposure to this product."

Some adverse effects in cattle

It is thought that long-term adverse health effects of chlorophenols are likely to be caused by its impurities, which include various dioxins. A 1980 study using cattle compared the toxicity of technical and analytical grades of PCP. The lower grade technical PCP, which contained more dioxins, produced health effects in cattle, including increased liver and lung weight, decreased thymus weight and abnormal functioning of the gall bladder. Except for minor symptoms, cattle receiving the higher grade analytical PCP (with less dioxins) were comparable to the unexposed control group of cattle.

PCP spills pollute the environment

PCP can enter the environment both intentionally and accidentally. Intentional disposal of PCP can occur in the manufacturing process, at wood treatment plants or during on-site treatments with wood preservatives. Accidental dispersal can occur from manufacturing mishaps, such as spills from ruptured processing equipment, or broken storage containers. Although these accidental spills contribute a small volume of PCP overall, they can be significant in the immediate vicinity of the accident. In 1972 in British Columbia, a hydro pole was treated in place with a PCP preservative. A nearby stream received a shock load of PCP, killing fish for a distance of 800 metres downstream.

Contamination of cattle from chewing fences

As already mentioned, chlorophenols can be accidentally mixed with food or ingested by animals. Some routine studies done by Agriculture Canada between 1984 and 1986 indicated that a significant percentage of all pork samples contain over 0.1 mg/kg of PCP. Lower levels of PCP were found in poultry products. Work completed in 1987 by the Ontario Pesticide Laboratory showed a correlation between residues of PCP in wood shavings and those found in poultry. The evidence suggests that using PCP-treated wood for such things as fence posts around livestock can result in significant contamination of the animals. Chlorinated water can also be a source of PCP. It has also been shown that chlorine can react with naturally occurring phenol to produce chlorophenols.

Data lacking on chlorophenol properties

PCP has the highest bioaccumulation potential of the chlorophenols, although even it accumulates to only 100 times its concentration in water in the worst situation. It is therefore not anticipated that there will be the same problem of bioaccumulation in the food chain as there is with some other chemicals such as DDT. On the other hand, the National Research Council warns that not enough basic data on the properties of chlorophenols exists to predict their environmental fate.

Evidence of bioaccumulation

In 1983, an accidental discharge of wood-impregnating solution containing chlorophenols in Sweden provided more information regarding longer term effects of these chemicals. The chlorophenols were rapidly distributed in the river system, and even after 2 months, they could still be detected in the livers of fish caught 15 km from the discharge point. Microbes converted the chlorophenols into compounds which bioaccumulated to a greater extent.

Traces of PCP found in urine

Chlorophenols and their impurities have been described as "probably ubiquitous" in the Canadian environment. A study in the US showed that the average American had 0.02 to 0.08 ppm of PCP in their urine samples; no such a study has been carried out in Canada. Trace amounts of PCP have been found in potatoes and other crops, and chlorophenols have been found in water and in sediment samples.

Alternative wood preservatives

It is estimated that approximately 3.6 billion board feet of British Columbia lumber worth $2.1 billion are treated with wood preservatives annually. The demand for very clean wood is on the rise so that the availability of wood preservatives is important. Several new alternatives to the chlorophenol chemicals are registered for use in Canada, one of them being borax. These alternatives do not work as effectively on the full range of organisms. They also do not have the proper health and environmental data bases required to meet current Canadian standards. The cost of employing these alternatives could run between $10 to $48 million resulting in a one to two percent increase in production costs. In Canada, employing wood preservatives like PCP is still cheaper than drying the wood in kilns.

Revision of standards for PCP

Products containing PCP and PCP salt have been registered for use in Canada since 1949. In 1981, Agriculture Canada, under the authority of the Pest Control Products Act, announced revisions to standards for the use of chlorophenols. A ban was put on several chlorophenol-containing products meant for wood preservation in the home or on the farm. Included in this ban are chlorophenol-containing wood preservatives for use on above-ground interior woodwork of farm buildings, such as on chicken roosts. There is a ban on chlorophenol-containing products labelled for a variety of uses such as herbicides, fungicides, and soil sterilants. As well, the regulations limit the use of these chemicals as additives to textiles. Products for use on textiles now must contain the warning: "Do not incorporate into materials of which end use will result in prolonged direct skin contact, eg life jackets, sleeping bags, sports equipment." Agriculture Canada decided to terminate all PCP sapstain control uses in Canada by 1991.

CONCLUDING REMARKS

Direct risks to human health are not high

In industrially developed countries like Canada, the actual risks to human health from pesticides in the environment are small relative to the risks from many other chemicals. The use of pesticides in agriculture, public health and home is closely regulated, and the users of pesticides are generally quite well informed regarding the risks. There is evidence that we are likely to be at considerably greater health risk from naturally occurring chemicals in foods than from residues of synthetic chemicals.

Pesticides residues are declining

As people became informed of the hazards of pesticides in the 1960s, the amount of pesticide residues in the environment has decreased markedly. Measurement of pesticide residues has become more accurate and sensitive. Stricter regulations have been introduced. Persistent broad spectrum pesticides have been replaced by more specific, less persistent chemicals.

Perceived risk exceeds the actual risk

The perceived risk in Canada of pesticides in the environment to human health seems to be significantly greater than the actual risk. Several phenomena may explain this. There is a time-lag in our perception; thus the serious environmental problems caused by pesticides some decades ago are still in the media and in our minds. The risks posed by pesticides in Third World tropical countries are significantly greater than in Canada, but these problems may be perceived to be universal. Any exposure to pesticides in the environment is also an involuntary exposure and, as explained in Chapter VIII, involuntary exposures are usually perceived to have a much greater risk than they actually have.

Less regulation and monitoring in Third World

The actual risk of pesticides to health is considerably greater in Third World countries for a number of reasons. Farming there is more often on a smaller scale. There are thus more individual farmers, many of whom lack the proper knowledge about safe, effective use of pesticides. Less regulation regarding the use of chemicals exists in these countries, and there is less monitoring and enforcement of the regulations that have been introduced. As well, there is great pressure to increase food production in any way possible in developing countries. Third World countries also tend to be in tropical regions, where the climate favours the pests which have an effect on agriculture and on public health. The risks of pesticides in the environment in Third World countries may however not seem that great there relative to the risks of communicable diseases, food shortages and poverty.

Human health risks from pesticides come usually from the greater exposures that occur in occupational use rather than from residues in the environment. Perhaps the greatest human health risk of pesticides is the availability of pesticides for use in suicide attempts. Another significant human health risk involving pesticides exists but it is not discussed in this book. That is the risk in the manufacture. The tragic accident in 1984 at Bhopal, India occurred in a pesticide plant. Over 3,000 people died from the emissions and many thousands became seriously ill. Again, there is a greater risk of such accidents in Third World countries because of fewer regulations, less monitoring and generally fewer resources for use in health and safety.

Accident at Bhopal pesticide plant

The environmental effects of pesticides can be serious. Several species of birds were threatened with extinction in North America because of pesticide build-up. Such threats to the existence of species still exist in tropical countries where persistent pesticides such as DDT are being used. This of course also affects species in the temperate zones because of migratory patterns. Rachel Carson in her book "Silent Spring" gave a very eloquent description of what nature would be like if the world were to rely on very large-scale use of pesticides.

Some species of birds at risk

Pesticides certainly affect the natural ecosystem. They are designed to favour some species (eg grains) over others (eg weeds), or to reduce some species (eg mosquitoes). Past experience indicates that humans would be wise to minimize their disturbance of the natural ecosystem. Thus a minimum of pesticide use is desirable; this is the aim of Integrated Pest Management. Less disturbance of the natural ecosystem should lead to greater sustainability of that system.

Ecology threatened by pesticides

It is as yet uncertain how much we can cut back on the current use of pesticides. These chemicals are used to better achieve human goals such as increased agricultural production, enhanced public health and aesthetically improved gardens. As the risks of pesticides have become more apparent, there have been more attempts to cut back on their use. Discontinuing the use of all pesticides may result in starvation if food production cannot be maintained, and in epidemics if disease-carrying insects multiply.

Problems in reducing pesticide use

Eliminating pesticides could also result in the destruction of forests and other natural ecosystems if more arable land is required to grow sufficient food. It is unlikely that completely "chemical free" agriculture can maintain the productivity per hectare of modern industrial agriculture without considerable extra human input. This may nevertheless be required if, as predicted by some scientists, our current chemically-dependent agriculture proves not to be sustainable. No certain answers to these problems are available today.

Prospects for food production unclear

Recommended Reading

"The Pesticide Handbook", 3rd ed. International Organization of Consumers Unions, Penang, Malaysia, 1991.

"Risk Assessment of Pesticides", A Chemical and Engineering News Forum. *Chemical and Engineering News*, pp 27-55, January 7, 1991.

"The Death of Ramon Gonzalez: The Modern Agricultural Dilemma", A. Wright. University of Texas Press, 1990.

"Silent Spring Revisited", G.J. Marco, R.M. Hollingworth and W. Durham, eds. American Chemical Society, Washington, D.C., 1987.

"Breaking the Pesticide Habit: Alternatives to 12 Hazardous Pesticides", T. Gips, International Alliance for Sustainable Agriculture, Minniapolis, Minn., 1987.

CHAPTER VII
METAL POLLUTION

Metal pollution is an old problem

Metal pollution is not new. Writers from ancient Rome documented illnesses suffered by workers in a Spanish mercury mine in the first century AD. Roman writers also observed that people became unhealthy from breathing fumes from metal mines and told of water pollution near mines. To protect their own immediate environment, the Romans prohibited mining in Italy, but nowhere else in their empire.

Humans have used metals for a long time

Metals are rooted as deeply in human history as archaeologists can dig. About 8,000 years ago, a group of craftsmen in western Asia made a discovery that would have far-reaching implications for the human race. They stumbled on lumps of almost pure gold and copper; more importantly, they had the curiosity to investigate these materials. These lumps of metal must have been particularly attractive to them, not only because of their rich lustre, but also because of their pliency. Unlike clay, copper and gold could be softened or liquified with intense heat, and then beaten into sheets or moulded into specific shapes. The early metal-smiths soon discovered that the same intense heat could be used to extract these metals from metal-bearing ores. This discovery of the process of smelting marked the beginning of the age of metals.

Metals give technological advantages

Successive epochs in human history bear the names of the metals which enabled the leaps in technologies that characterized those times. By 3000 BC, metal-smiths had learned that combining a little tin with copper resulted in a metal that was harder than either of its components. With this discovery, the Bronze age was born; an age marked by vast improvements in the quality of metal tools. Then came iron, harder and tougher than bronze. Knowledge of this wonder-metal had spread from Asia all across Europe by 800 BC - the age of Bronze had given way to the Iron age. Throughout these eras, many new metals were discovered in addition to bronze and iron. For instance, in their search for silver, ancient metalworkers also found lead. Silver is often deposited in lead ores; after the lead was extracted from its ore, the silver could be recovered. Lead was found to be a useful metal in its own right; the Egyptians used lead-glazed pottery, the Greeks had the first lead-covered roofs, and the Romans used lead pipes for their plumbing systems.

Metals have shaped human history

The ancient metals were instrumental in shaping the course of human history. Those societies who seriously explored the possibilities of metals found themselves at a great advantage. Metal tools facilitated the development of farming and the formation of armies. They even allowed vast improvements in an earlier invention, the wheel. With metal knives, wooden spokes could be cut and shaped so that the wheel became lighter and more versatile. The ways of using and working metal eventually spread to nearly all peoples, but those who were innovative often enjoyed the power afforded by exclusive knowledge for long periods of time.

The earth's crust contains many metals

Not surprisingly, gold, copper, silver, tin, iron, lead and zinc are often referred to as "the metals of antiquity". Even now, the word "metals" usually brings to mind these historic elements. Actually, there are over 50 metals and semi-metals on our planet, all of which are mined from ores in the earth's crust.

We have found uses for nearly every metal

Gradually, as the techniques of mining and smelting grew more refined, more and more metals became accessible to societies. Today, nearly all of the metals found on earth have found some application in industry. Humans have derived great benefits from exploiting the properties of metals. But in this exploitation of metals, we unwittingly embraced the consequences of metal pollution.

Excess intake of all metals is harmful

It is essential for good health that we ingest small quantities of a number of metals. Two major ones are the calcium needed for our bones and the iron for our blood. In larger doses, even these seemingly safe substances become poisonous to our systems. Ancient miners and metal-smiths did not have such information, and many of them paid for their ignorance with their health. The mines and smelters were operated on a small scale, but their uncontrolled techniques often resulted in severe local contamination.

Ignorance of hazards caused harm

Many metals were also ingredients in compounds to which people were directly exposed. Compounds of mercury and lead were used as medicines to treat ailments such as diarrhoea and syphilis. They are still ingredients in some soaps and cosmetics. We can probably thank mercury for the expression, "Mad as a Hatter". Workers in the felt hat industry used mercury solutions to treat fur; they were sometimes poisoned by the mercury they handled, suffering what was called "hatter's shakes". With continued exposure, some of them went mad - perhaps inspiring Lewis Carroll's "Mad Hatter" of Alice in Wonderland.

Did lead cause the fall of the Roman Empire?

Many scientists and historians now believe that whole societies may have suffered from toxic metal effects. It has been suggested that lead was a critical influence in the fall of the Roman Empire. The Romans used a wide variety of lead products in their houses and kitchens, and many researchers now agree that some Romans must have suffered from sickness and insanity due to lead poisoning. Ironically, the Roman ruling classes were most at risk because they could afford the lead pans, pipes, paints and pewter that the poor could not. These same innovative items that afforded the Roman elite enviable comforts constantly burdened their bodies with small doses of lead.

Lead compound used as sweetener

This situation was probably exacerbated by the Romans' apparent sweet tooth. They traditionally added boiled-down fruit syrups, called sapa, to their wines - and they drank a lot of wine. Old recipes for sapa required that fruit juices be boiled in a lead pan. These recipes produced very sweet syrups because acids in the juice react with lead to form lead acetate, a substance as sweet as sugar. Recipes for sapa produced syrups that contained over 1000 parts per million (ppm) of dissolved lead, compared to natural lead levels of around 0.001 ppm found in the environment and living things. Bone analysis lends support to the theory of lead poisoning in ancient Rome. The level of lead in Roman bones has been found to be as high as 2000 ppm - far in excess of the average 30 ppm in the bones of a modern North American.

Whether or not lead played a part in undermining the Roman empire will always be controversial. It is clear, however, that all of the historic cases of metal pollution occurred on a local or regional scale. Since mining, smelting and manufacturing took place in small, isolated facilities, the consequences of metal pollution were borne mainly by those who were in direct contact with the polluting products or processes.

Early metal pollution was on local scale

The industrial revolution changed the nature of metal pollution forever. In the 19th century, large parts of Europe were transformed from agricultural to industrial societies. The factory system came into being. Mining and manufacturing operations took on enormous proportions. The new, industrialized age also stimulated a virtually endless demand for coal, iron and steel. Huge industrial complexes with imposing smoke stacks changed the face of Europe. As the demands for manufactured goods grew larger, so did the quantities of waste produced. The factories disposed of these wastes on a grand scale, flushing effluent into rivers and sending smoke and ash through stacks into the air. Metal pollution, which had once been a local phenomenon, now became global in scale.

Factory system changed scale of pollution

PROPERTIES AND USES OF METALS

Today, there are probably no known metals that have not found some application in our industrial society. As technology continues to flourish, the numbers of uses for metals will likely continue to grow. It is no accident that metals are an integral part of our daily lives. They share a number of chemical and physical properties that are particularly useful to us. Individual metals have many additional specific properties, but the most widespread uses for metals are based on the qualities of malleability, strength, lustre and electrical conductivity.

Metals possess useful properties

Nearly every use that metals have been put to depends on the fact that all metals are, to some degree, malleable. That is, they can be melted and cast in molds, heated and pounded into thin sheets or drawn into fine wires. Whatever the method, a metal that has been molded will remain that way unless it is subjected to a high level of stress.

Metals can be molded

The potential of metals is further expanded by mixing metals with other elements to make alloys with additional desired qualities. Steel results from combining iron and carbon. The strength and resilience of steel makes it the nuts and bolts of the construction industry - literally. Without steel girders, the skyscrapers that pervade modern cities would still be an architect's dream. The addition of nickel to steel gives it corrosion resistance; that alloy is known to us as stainless steel. More recently, new combinations of metals have been discovered which match steel for hardness and strength but are much lighter. For instance, an alloy of copper, magnesium, manganese and aluminum called Duralumin is now the most widely used substance for aircraft construction.

Mixtures of metals improve properties

Lustre is a valued property

Another commonly used alloy is brass, a mixture of copper and zinc. It is extensively used for structural products with ornamental qualities. When brass is chosen to make doorknobs, light fixtures and hinges, it is not only because of its strength and hardness, but also because it is lustrous. Chromium is another example of a metal that is put to widespread use based largely on its aesthetic properties. This highly lustrous, non-tarnishing silver-grey metal is used to plate all kinds of objects. Like our ancestors, we find the lustre of precious metals such as gold and silver so captivating that they continue to be highly cherished. Indeed, these two rare metals are so valuable to our society that they are used as a store of wealth.

Metals used to conduct electricity

Gold is also extremely useful to the electronics industry because it is an excellent conductor of electricity and is non-corrosive. Very thin gold plating is commonly used on electronic contacts, like those found in computers, where it is important to avoid any corrosion which could impair the signal. But the more familiar use for highly conductive metals is as wiring for electrical transmission. Copper and aluminum are the prevailing choices for wiring since they are also good conductors with the added advantage of being fairly cheap.

Crustal abundance of metals

Metal	Abundance (ppm by mass)
Aluminum	83,000
Iron	58,000
Magnesium	28,000
Manganese	1,300
Vanadium	170
Chromium	96
Zinc	82
Nickel	72
Copper	58
Lead	12
Arsenic	2
Tin	1.7
Uranium	1.6
Silver	.08
Mercury	.08
Gold	.002

Use of metals involves hazards

This perusal of the general properties and uses of metals illustrates the enormous benefits accrued by society from them. However, this relationship with metals does not come without risk. Through exposure to metal pollution, people and the environment have suffered the toxic effects of excessive amounts of metals. Particularly prominent are lead, mercury, cadmium and arsenic. Because of the controversy that has often surrounded them, the uses of these historically important metals will be examined in more detail. These are by no means the most common metals in the earth's crust, as shown in the above table. The two most common metals are ones which we use in large amounts and for many purposes in our society: aluminum, which accounts for 8.3% of the crust, and iron, which accounts for 5.3% by weight.

Lead

Lead is a soft, malleable, stable, heavy, bluish-grey metal. Canadian industry consumes half of the roughly 1.8 million tonnes of lead produced annually in Canada. In Canada, the largest single use for lead is in the manufacture of lead-acid storage batteries for automobiles. Because of its corrosion resistance, lead is also used in semi-finished products, such as pipes, sheets, foils and ammunition. Until 1930, lead-based pigment was widely used as the chief whitening agent in paints. This was eventually found to be a major source of lead poisoning. About one-quarter of the lead produced annually is used in the manufacture of chemicals. The most important of these was tetraethyl lead, an anti-knock additive for gasoline. The recent banning of leaded gasoline in North America has effectively eliminated this use of lead and drastically reduced single largest source of global lead pollution.

Lead used in car batteries, pipes

Mercury

Mercury is relatively rare in nature, but has been mined since Roman times. It combines with many other metals like tin, zinc, and copper to form "amalgams". Such amalgams have been used by dentists to fill dental cavities. Mercury also combines with organic elements, such as chlorine, but is most commonly found in compounds such as mercuric sulphide (HgS), the brilliant red mineral called cinnabar. Pure mercury is usually obtained by extracting it from the cinnabar with heat. The mercury in the cinnabar evaporates, and can then be condensed and collected.

Mercury forms amalgams

Mercury has certain physical and chemical properties which make it uniquely attractive for thousands of uses. Because mercury is toxic to bacteria, fungi, and other organisms, organic mercury compounds are used in slimicides and fungicides. For instance, in the pulp and paper industry, organic mercury compounds have been used to kill bacteria, fungi, and yeast-like organisms. In latex paints, mercury has been used to inhibit the growth of mildew. In agriculture, mercury-containing fungicides are applied to leaves, seeds and soil.

Use of mercury in pesticides

The most familiar use for pure mercury is in thermometers, barometers and manometers. This use takes advantage of two of mercury's more unusual characteristics. It is one of the very few metals which is a liquid at room temperature and which remains liquid over a wide temperature range. It has a freezing point lower than any other metal (-39°C). Mercury also expands uniformly as temperature increases, making accurate measurements possible.

Mercury is a liquid at normal temperatures

Many important uses of mercury depend upon its electrical properties. Large amounts of the metal are used in electrical products such as switches, batteries, and mercury vapour lamps. Most of this mercury is recovered and reused, but some escapes to the environment from accidental breakage. In the case of mercury vapour lamps, the problem of mercury pollution through breakage has led to a general shift to sodium vapour lamps as an environmentally safer alternative for street lighting.

Mercury use in lamps declining

Mercury no longer used for chlor-alkali

Mercury was also widely used until the mid-1970s in the chlor-alkali industry. In "mercury cells", this metal functioned as an efficient cathode for the electrolysis of sodium chloride salt to produce chlorine. Unfortunately, the mercury leaking out of the cells was entering wastewater and polluting rivers and lakes. The electrolytic system has largely been replaced now by the alternative "diaphragm cell" system.

Arsenic

Arsenic is not a true metal

Arsenic is actually a semi-metal. It is a silver-white element which differs from true metals by being brittle rather than malleable. It is present in the earth's crust with an average concentration of 2 ppm, usually in association with lead and gold ores, and in coal.

Ideal as a poison for centuries

Arsenic is probably best known for its poisonous properties; less than 100 milligrams of arsenic trioxide (or white arsenic) can be enough to kill a person (see Chapter VIII). Indeed, from the Middle Ages until the end of the nineteenth century, it was considered the perfect poison: many an unwanted lover has joined the ranks of the kings and popes that have been disposed of by arsenic poisoning. Arsenic's popularity for murderous deeds is not based solely on its toxicity, since many organic chemicals are far more poisonous. Arsenic poisoning had, up until recently, the advantage of being undetectable. Not only were the symptoms of arsenic poisoning easy to mistake for other ailments, but the compound remained difficult to detect in body tissues, food or drink until the mid-1800s.

Arsenic used in pesticides

Today arsenic's toxicity is used to poison insects and weeds. The largest single use for arsenic is in making agricultural pesticides such as calcium arsenate and lead arsenate. The herbicide sodium arsenite was widely used as a pre-harvest defoliant for potato plants until its use was phased out in 1971. At times, other chemicals have been sought out as substitutes for the more widely used arsenic pesticides. For example, DDT was a popular alternative until it was banned (see Chapter VI).

Traces of arsenic are essential for life

It may seem strange, but while some farmers are spraying arsenates on their crops to kill pests, others are deliberately using it to fatten their pigs and poultry. Apparently, arsenic is one of the essential elements; we need small amounts to sustain life. Furthermore, small doses of arsenic have proven to be very good at promoting growth, and a compound called roxarsone is now widely used as a growth stimulant, particularly in the US. This is allowed because the animals excrete the arsenic very quickly, so that if the farmer stops doping the feed a few days before slaughter, the arsenic levels fall to legally acceptable levels of one ppm or less.

Arsenic useful in alloys

Arsenic is also used as a hardener in the making of alloys. The addition of up to three percent arsenic hardens lead; arsenic is used to this end in products including lead-based bearing alloys for internal combustion engines or lead-based battery plates. The addition of small amounts (around 0.1%) of arsenic to copper can improve corrosion and erosion resistance. Arsenic also has miscellaneous uses such as hide tanning, the manufacture of paint pigments and as an ingredient in wood preserving compounds.

Arsenic in electronic materials

In the last few years, very pure arsenic has been in demand by the electronics industry for use in making semi-conductors. As gallium or indium arsenide, arsenic can be found in transistors, in lasers that operate at room temperatures, and in infra-red devices. It is also added to elements like germanium and silicon to confer semi-conductive properties.

Cadmium

Cadmium a by-product of zinc production

Cadmium commonly occurs in nature combined with sulphur. Such cadmium sulphide minerals are usually found with zinc deposits, as the chemical properties of cadmium and zinc are similar. Because of this close association with zinc, industrial grade cadmium is usually obtained as a by-product of zinc refining. Canada produces an estimated 1000 tonnes of cadmium metal in this manner. Nearly an equal amount of cadmium is present as an impurity in the zinc concentrate produced.

World cadmium production rising

In the latter half of this century, the world's production of cadmium has quadrupled; Japan, the US and the Soviet Union are the leading producers. Canada exports some cadmium selenite which is used in photoconductors, semi-conductors and ceramics. But overall, we are net importers of cadmium in that we import many of the various products that contain this metal.

Cadmium is an anti-rust coating

Cadmium compounds are used in pigments and plastic stabilizers; thus some ceramics, porcelain and tinted plastics contain this metal. One rapidly growing use for cadmium is in nickel-cadmium batteries, such as those used in pocket calculators. Cadmium is effective as an anti-rust coating on other metals, and can be found in televisions, radios, in automobile components like radiators, or in coffee percolators. The corrosion resistance of cadmium is particularly useful for electroplating steel that will be used in hot, humid or marine environments. In fact, the plating of steel for marine uses accounts for 75% of industrial cadmium use in Canada.

ENTRY INTO THE ENVIRONMENT

Great increase in metal production

Since the turn of the century, levels of metal consumption in industrialized nations have climbed sharply. Between 1930 and 1985, worldwide mine production of zinc increased by a factor of 4, copper by 5, chromium by 18, nickel by 35 and aluminum by a factor of 114. The rapid rise in the use and manufacture of metallic products in the 20th century is related to a considerable extent to the growth in population, but also to an increase in the standard of living and to technological developments.

Variety of products for consumers

The material standard of living in the industrialized world has risen considerably since the 20th century began, in particular since World War II. As it becomes easier to secure the necessities of life, more and more people are able to look beyond the bare essentials, to achieve levels of convenience and comfort in their lives that their parents could not even have imagined. The burgeoning technology of this century has enabled the development of a vast range of products which act as both stimulus for and response to growing consumer demands.

Pollution follows consumption

This cycle of increasing consumption involves another, less obvious reality which is by no means exclusive to the issue of metal pollution. In a society that still accepts the notion of "disposable goods" and permits wasteful manufacturing practices, every use inevitably creates waste - waste that is discarded to the environment. Pollution grows as production grows.

Metals do not biodegrade

Metals are elements, and are thus non-degradable. They cannot be eliminated through incineration or biological processes. Consequently, metals can be cumulative pollutants; they can accumulate in ecosystems, and often, in food chains. Some metals, including lead, cadmium and mercury, have no known beneficial health effects. Even those which are essential to life in certain quantities, such as magnesium or zinc, are toxic when present in excess. As metals are mobilized from their stable forms in ore minerals by mining, they accumulate over time in more active forms in the environment. They can thus be expected to cause progressively more serious health effects in the organisms exposed to them.

Humans upset biogeochemical cycles

Natural processes, like weathering and erosion, remove metals from the bedrock and keep them circulating through air, water, soils and all living matter. If undisturbed by human activity, the natural biogeochemical cycle ensures that the distribution of any given metal within an ecosystem is held roughly constant. The increase in metal use, primarily by the industrialized countries of the northern hemisphere, has upset this natural balance. The quantities of most trace metals introduced to the environment through human activities now far outweigh natural sources, and our actions may have overwhelmed the natural biogeochemical cycle of trace metals in many ecosystems.

Industrial inputs exceed natural

A recently published inventory of global atmospheric emissions of trace metals show that industrial emissions of lead exceed natural inputs by a factor of 28, of cadmium by 6, and of vanadium and zinc by a factor of 3. Similarly, human activities are responsible for introducing considerably more arsenic and mercury to the environment than natural processes. Atmospheric metal pollution arises mainly from the mining, smelting and refining of metallic ores; the manufacture and use of metallic products; and the burning of fossil fuels. Smelting is the leading source of atmospheric copper and zinc, while the manufacture of steel is the greatest source for chromium and manganese. The burning of fossil fuels to generate energy is responsible for over 95% of the vanadium and 80% of the nickel released to the air worldwide.

Airborne metals often represent the largest input of pollutant metals into water. The atmosphere can be responsible for having transported most of the trace metals detected in many aquatic systems. Over 50% of all of the trace metals found in the Great Lakes are deposited through atmospheric transfer. Freshwater ecosystems are very sensitive to additions of toxic metals. Since natural background levels of metals are usually very low in freshwater, even small additions of some metals can have a significant impact on the ecology.

Air transports metals to water

The oceans are also affected by atmospheric deposition of toxic metals. Evidence is difficult to gather because of the enormous volumes involved in the marine cycling of metals. But studies involving lead reveal that atmospheric inputs of that metal are altering the lead signature of certain oceanic regions. In the North Atlantic, recently deposited coral shells contain 15 times more lead than layers put down a century ago.

Oceans are also polluted by fallout

As we have seen throughout this book, urban areas tend to be ecological "hot-spots" of pollution, and the case for metals is no exception. In waters like the Great Lakes, point source emissions such as domestic and industrial waste effluents and the dumping of sewage sludge combine with atmospheric fallout to create large local burdens of toxic metals - in water as well as air. The levels of toxic metals found in the rainwater of some urban areas already exceed levels considered safe for human consumption. This can have serious implications in developing countries where people often rely heavily on rainwater for drinking, and where rapid industrialization can give rise to increased atmospheric emissions of metals.

Metal pollution greater in urban areas

In urban hot-spots, air and water are not the only environmental compartments carrying elevated levels of metals. Soils and sediments are the ultimate sink for many pollutants - trace metals included. We dump enormous quantities of waste onto land, and the input from such garbage disposal to metal pollution of soils can be much greater than that from atmospheric fallout. Fifty-five to 80% of the metal pollution in soils is thought to be derived from the decomposition of metal-containing products and the disposal of coal and wood ash.

Urban soils can be heavily polluted

In rural areas, atmospheric deposition of metals plays a larger role in soil pollution. Arable soils also receive significant metal burdens from pesticides, fertilizers and animal waste. Unfortunately, metals tend to accumulate in the surface layers of soil, which are the most biologically active regions. This means that metals are often readily available for uptake by crops and vegetables. If the loading rates for metals continue to rise, many arable soils will become unable to produce edible food. In Japan, this point has already been reached: excessive metal contamination has rendered 9.5% of all rice paddy soils unsuitable for growing rice fit for human consumption.

Excess metals make soil unusable

The trends in metal pollution discussed so far indicate that it has the potential to become a serious problem. This is particularly true for the "big four": lead, mercury, arsenic and cadmium. Accordingly, we shall now turn to an examination of the specific sources of emissions for these metals.

Metal pollution is a growing problem

Lead

Car exhausts a large source of lead pollution

In terms of air pollution, lead emissions are an exception to the typical sources of atmospheric metals mentioned earlier. The use of leaded gasoline in automobiles is still the single largest global source of airborne lead pollution. By 1990, some European countries, Canada, and the United States had phased out the use of leaded gasoline. The move was partly in response to scientific evidence from a study in Italy which found that exposure to lead emissions from gasoline accounted for at least 24% of people's blood lead levels. It was also spurred by efforts to control emissions of automotive exhausts responsible for the formation of smog. As discussed in Chapter I: Outdoor Air Pollution, catalytic converters cannot be used if lead is present in gasoline. However, despite these reductions in the use of leaded gasoline, automobile exhaust gases still account for about 60% of all lead emissions worldwide.

Pollution greatest near point sources

Other sources of atmospheric lead do follow the general patterns for metal emissions. The mining, smelting and refining of lead ores and the production of primary copper and nickel all introduce lead to the air. The burning of coal releases about 3,500 tonnes of lead to the world's atmosphere every year. In all of these cases, including emissions from gasoline, the atmospheric lead burdens that result are especially heavy near the actual sources, whether that source is a smelter or a downtown traffic jam. In polluted areas, both plants and animals are susceptible to the harmful effects of lead. Through their leaves, plants can absorb lead directly from the air, and animals can absorb it through their respiratory systems.

Most water pollution comes from air

Atmospheric transfer of lead is also the major pathway for the pollution of water. Over 70% of the lead found in aquatic systems worldwide is deposited from the air. High levels of lead pollution threaten aquatic ecosystems both through accumulation within the food chain and direct uptake by aquatic animals. Fish can absorb lead both through the food they eat and directly from the water through the skin and gills. Of course, lead pollution in water also threatens people. In Canada, drinking water standards have been set at 50 micrograms of lead per litre of water (ug/L) but there is now some debate as to whether this limit should be lowered.

Municipal waste a source of lead

Lead-containing products are widely found in municipal wastes. Television sets, radios, glass, ceramics and plastics all contribute small amounts of lead to the environment when disposed of in garbage dumps. The highest contribution to lead pollution comes from the lead-acid batteries commonly used in automobiles. In the US, over 80% of these batteries are now recycled, but enough are still thrown into garbage dumps to make up about 65% of the lead found in municipal waste.

Mercury

Mercury released in combustion

Worldwide, human activities contribute about 20,000 tonnes of mercury to the environment every year; this compares to an estimated 3,000 tonnes from natural processes. The atmospheric emissions from burning fossil fuels, especially coal, account for about 60% of this total. Like many trace metals, mercury is also released to the air during the mining and refining of the mercury ore. Incineration of garbage containing discarded electrical equipment is becoming an increasingly important source of atmospheric mercury.

Mercury in water from industrial waste

Atmospheric transfer is not the principal pathway for mercury found in waterways, although it does account for 30% of aquatic mercury pollution globally. Most mercury is directly emitted to waters through industrial discharges of trace mercury in waste effluent. In particular, the chlor-alkali and pulp and paper industries have been responsible for serious local contamination of many waters.

Steps for reduction of mercury

Discharges from the chlor-alkali industries have either ceased or decreased markedly as a result of either switching to the "diaphragm cell" process, or recirculating waste water and installing lagoons and settling ponds in which the mercury collects. The pulp and paper industries have either discontinued the use of organo-mercury compounds or have installed waste treatment procedures to limit the amount of mercury released. In spite of the large reductions in mercury emissions by many industries, their legacy of mercury pollution will last for years to come.

Organic mercury is more harmful

Mercury released to waters tends to bind to particles of dirt and settle to the bottom of lakes and rivers. This is not necessarily harmful to the aquatic ecosystem, until the mercury is altered by bacteria in the water to form methylmercury. This form of mercury dissolves in water, is easily absorbed by aquatic animals, and can accumulate in tissue until it reaches toxic concentrations. Consequently, the effects of mercury pollution on aquatic life can continue 20 to 30 years after the mercury emissions have been halted.

Flooding of land mobilizes mercury

One recently discovered source of mercury pollution in water has the potential to become a very contentious issue in Canada. Hydroelectric dams may be responsible for significant mercury pollution in surrounding waters. Some such dams flood large tracts of land, creating enormous reservoirs. It seems that during the breakdown of the newly submerged vegetation and soils, significant quantities of natural mercury are liberated to the water.

Mercury pollution at James Bay project

For the gigantic James Bay hydroelectric project in Quebec, mercury pollution of the reservoirs is a major issue. A ten year study published in 1988 by Hydro-Quebec confirmed what the Cree people of Northern Quebec had alleged for years. Fish samples taken from the reservoirs contained concentrations of mercury nine times the Federal limit deemed safe for human consumption. This pollution has profound effects on the James Bay Cree. Fresh fish make up a large part of their diet and they have now been advised to stop eating fish from waters that are not expected to return to acceptable mercury levels for 30 years.

Arsenic

We mobilize more arsenic than nature

Approximately 40,000 tonnes of arsenic are committed to the global environment by natural weathering processes every year; human activities now release twice this amount. The principal industrial sources of atmospheric arsenic are similar to those for most metals: coal combustion and metal smelting and roasting (primarily of gold, copper and nickel) are two major sources of arsenic. A large fraction of airborne arsenic is also derived from the spraying of arsenic pesticides.

Plants absorb arsenic from the soil

The use of arsenic pesticides greatly increases the arsenic burden in many agricultural soils. Crops and vegetables planted on soils with high concentrations of arsenic will readily absorb increased amounts of arsenic. Consequently, some foods contain elevated levels of arsenic.

Arsenic enters water in many ways

Soil run-off is an important source of water pollution, but so are industrial discharges from mining and refining processes and related chemical industries. Indeed, a large fraction of the world's production of arsenic - roughly 50,000 tonnes annually - reaches rivers, lakes and oceans. Not surprisingly, contaminated drinking water, both surface and groundwater, is an important route for human exposure to arsenic in many urban and heavily cultivated agricultural areas.

Standards for arsenic vary

Allowable exposure levels for arsenic vary considerably. In the US, the Occupational Safety and Health Administration (OSHA) standard for arsenic in air is 0.010 milligram per cubic metre (mg/m^3). In Canada, the federally recommended Threshold Level Value (TLV) is 0.20 mg/m^3. Two provinces have set lower levels. British Columbia, at 0.050 and Alberta at 0.050 with a 0.150 peak. Ontario follows the 0.20 guideline and Quebec allows 0.500 mg/m^3. The Canadian Drinking Water Standard for arsenic sets the maximum permissible arsenic level at 50 ug/L. Maximum allowable arsenic limits in food (set by the Health Protection Branch, Health and Welfare Canada) are two ppm for fruits and one ppm for fresh vegetables.

Cadmium

Smelters are a source of cadmium in air

Cadmium is similar to arsenic in that it is usually emitted as a by-product of various industrial activities. The major sources of atmospheric cadmium are emissions from zinc, lead and copper smelters. Small quantities of cadmium can also be released during the burning of fossil fuels. But the cadmium content of coals and petroleum is so low that this is considered a very minor source of airborne cadmium.

Industries emit cadmium to water

Industries involved in the manufacture of alloys, paints, batteries and plastics are important sources of trace cadmium to both air and water. Direct industrial discharge and runoff play a larger role in cadmium water pollution than does atmospheric transport. The atmospheric pathway accounts for only 20% of the cadmium entering waters worldwide.

Soils are polluted with cadmium in several ways. In addition to atmospheric deposition, soils receive cadmium doses from the decomposition of cadmium-containing products in garbage dumps. Certain fertilizers and pesticides also contribute traces of cadmium to agricultural soils.

Fertilizers, garbage pollute soil

HEALTH EFFECTS

People have been aware of the toxicity of certain metals for centuries. Usually, concern about metal pollution has been raised by specific cases of gross, local contamination which had obvious consequences for human health. As a result, these rare case histories have been well studied, providing valuable insight on the acute toxicity of many metals. However, the effects of prolonged exposure to small doses of metals on humans, animals or ecosystems are still poorly understood. Yet chronic effects are the kind that most people in the world are more likely to experience.

Chronic health effects largely unknown

There are many reasons for this paucity of research into the chronic effects of environmental metal pollution. The usual experimental routes for verification of laboratory work are epidemiological studies (see Chapter VIII). These require medical data, which are difficult and expensive to obtain, taken over many years. Even if such information were available, the preconceptions and biases of the researcher can subtly influence the interpretation of the results. It is difficult to separate the effect of exposure to a metal from numerous other possible effects such as smoking. These influences, combined with the effects of differing experimental protocols, mean that results of different epidemiological studies are often conflicting and, hence, widely debated.

Conflicting results confuse the issues

Investigations of chronic metal poisoning are made even more difficult because the symptoms of metal toxicity in its early stages are non-specific. There are precious few identifiable early-warning signals. Often the diagnosable symptoms of metal poisoning do not become clear until the metal has accumulated in the body to reach a toxic dose. The early indications of effects of small doses of metals are at the cellular level: a general degradation of the health of cells, tissues or organs. The symptoms of such subtle damage either go unnoticed, or are symptoms shared by many other ailments. It may also be that this steady stress weakens the body, reducing its ability to fend off other diseases. In this case, the secondary disorder would probably be recognized, but the root of the problem would remain undiscovered.

Often no warning of metal poisoning

Some metals, lead and cadmium in particular, have long been known to be generally toxic to the heart. Many scientists now believe that these metals aggravate high blood pressure. Epidemiological studies have consistently found strong links between hypertension and elevated body levels of cadmium or lead.

Some metals linked to high blood pressure

Many metals also seem to be involved in causing cancer - at least in rats. Eighteen metals, including aluminum, arsenic, cadmium, chromium, lead, nickel and zinc are initiators or promoters of cancers in laboratory animals. So far,

Some metals may be carcinogenic

almost all of the work done to assess metals' carcinogenicity to people have been based on cases of very high exposure from accidental contamination of specific areas like the workplace or a local spill. There is much work to be done to assess what the risks are to the general population from exposure to small, environmental doses of these metals. Small doses are likely to lead to only small effects which are difficult to even detect. Nevertheless, small doses to a very large population can have a significant overall effect.

Reproductive disorders due to metals

Some scientists argue that certain metals must also be considered to be important environmental factors in the development of many reproductive disorders in humans. For example, lead has been implicated as a cause of a variety of pregnancy problems including spontaneous abortion, early membrane rupture and low birth weight. In men, lead has been shown to cause several reproductive problems such as dysfunctional sperm or low sperm count. Other metals, such as arsenic, are especially toxic to the foetus. Over the centuries, many women have used small quantities of arsenic to abort unwanted pregnancies.

Effects of metals on wildlife

Of course, animals are also at risk from elevated levels of trace metals in the environment. Urban animals are as affected by metal pollution as their human companions. Large numbers of cats, dogs and pigeons in cities are known to suffer from lead poisoning. In North America, biologists estimate that 2 to 3 million waterfowl suffer chronic lead poisoning, or what is called fatal plumbism, caused by eating spent lead pellets from hunters' rifles. Deer, moose and bear kidney in some regions of Canada contain very high concentrations of cadmium. Provincial officials in Ontario have issued advisories against eating the kidney meat, but the effects of these high levels of cadmium on the animals themselves are unknown.

Animals at top of food chain suffer most

Aquatic animals are very sensitive to metal pollution in the environment. Since metals tend to accumulate in living tissues, the effects of chronic metal poisoning are usually most visible in animals at the top of the food chain. The most familiar examples of this are the cases of mercury-contaminated fish in many waters throughout the world.

Metal pollution needs more study

There is still much to learn about the impacts of environmental metal pollution. Furthermore, the significance of what is known is subject to much debate among scientists and officials. The evidence gathered to date, especially for lead, mercury, arsenic and cadmium pollution, suggests that the health risks posed by environmental metal pollution warrant public attention.

Lead

Lead is everywhere

Lead is one of the most pervasive metals being released to the environment. Thanks largely to leaded gasoline and paints, nearly everyone has some lead in their tissues, even in remote corners of the world such as Nepal and the high Arctic.

Lead poisoning is associated with a wide variety of neurological and metabolic disorders. Consequently, the symptoms of acute lead poisoning are very varied. Early symptoms include stomach ache, weakness, irritability and fatigue. In later stages, victims may experience headaches, loss of appetite, pallor, drowsiness, vomiting, cramps, loss of coordination, convulsions and stupor.

Various symptoms in lead poisoning

Children are at particular risk from lead poisoning, partly because the effects of lead are proportionately greater in their smaller bodies, and partly because they touch and taste almost everything around them. The early years of life are also characterized by rapid brain growth. As a result, the major risk of lead poisoning to children is damage to their brain development. Acute lead poisoning can cause mental retardation in young children. Moderately elevated lead levels in children are associated with symptoms such as hyperactivity, poor hand-eye coordination, poor reading ability and generally decreased intellectual ability. Unfortunately, recent research by American scientists indicates that the neurological damage may be permanent.

Lead can impair brain function in children

In adults, lead can accumulate in the bones. As people age, bones dissolve, high levels of lead re-enter the blood stream and attack soft tissues. The liver is damaged, the kidneys may become irreversibly diseased, the heart may be affected, the lungs less able to rid themselves of dust and infectious agents, or the reproductive system impaired. Pregnant women are at particular risk from lead poisoning. In addition to any harm they sustain, lead crosses the placenta easily and can severely damage the brain development of the foetus.

Lead released from bones can harm adults

The re-circulation of lead transferred from bones is also linked to osteoporosis, or brittle bone disease, which afflicts elderly women. Since circulating lead interferes with the body's use of calcium, scientists think that accumulations of lead stored in bones may increase the chances of developing osteoporosis in later years.

Lead aggravates osteoporosis

The impacts of environmental lead poisoning on the general population are difficult to ascertain. But some scientists estimate that between 130-200 million people in the world carry blood lead burdens in excess of the suggested medical threshold of 200 ug/L. A growing number of medical researchers are of the opinion that there is no minimum amount of lead below which it causes no damage. If that is true, they say that the number of people exposed to potentially hazardous levels of lead must run to over one billion.

Lead may affect people worldwide

Mercury

Mercury's toxicity depends on whether it is in an organic or inorganic form. Metallic mercury, the kind found in thermometers, is an example of inorganic mercury. Mercury is called organic when it is combined with the chemical forms found in living matter (which are almost always based on carbon). Methylmercury is the most common form of organic mercury associated with human illness.

Two different toxic forms of mercury

Kidney, nervous system can be damaged

Inorganic mercury poisoning is primarily an occupational disease. Metallic mercury vaporizes easily and diffuses through the lungs into the blood, then into the brain causing serious damage to the central nervous system. Eventually, mercury tends to concentrate in the kidneys. After prolonged exposure, death may occur due to failure of the kidneys to function properly. From 1955 to 1975, the Workmen's Compensation Board of Ontario paid compensation to 24 workers affected by mercury poisoning. They were involved in such things as hat manufacturing, gold refining, fungicides, a dental laboratory, battery manufacturing, the electrical industry, and a chlor-alkali plant.

Symptoms of organic mercury poisoning

In terms of the health risk from environmental mercury pollution, organic mercury is far more important. The symptoms of acute methylmercury poisoning progress from numbness, hearing difficulties, speech and visual impairment, loss of motor coordination to paralysis, deformity, coma and death. As with many metals, the early symptoms of poisoning can be easily misdiagnosed as other ailments. In the 1950s, however, the people of Japan became terribly familiar with the symptoms of mercury poisoning.

Minamata disease in Japan

In 1932, an industrial chemical and fertilizer company called Chisso began dumping mercury compounds into Minamata Bay. This led to the accumulation of mercury in fish, a basic food item for the local people. Within a few years, local cats who were fed scraps of fish from the contaminated bay began jumping and twitching or running around in circles until they finally threw themselves into the water to drown. This disorder became known as Minamata or "cat dancing" disease. By the 1950s, people also began to suffer similar symptoms. In 1965, Minamata disease recurred in Niigata, Japan - again caused by consumption of contaminated fish.

Severe health effects in Minamata

Since that time, thousands of people have suffered brain damage, paralysis, or loss of hearing or sight due to mercury poisoning from Minamata Bay; at least 143 people died. The Japanese government has recently been ruled negligent in not acting to stop the spread of Minamata disease. In 1988, two top executives of Chisso were sentenced by the Japanese Supreme Court to suspended two-year prison terms on charges of professional negligence leading to death.

Incidences of methylmercury poisoning

The Minamata incident raised the world's awareness of mercury as an environmental hazard, and several subsequent accidents reinforced this. In 1956 and again in 1960, accidental methylmercury poisoning in Iraq resulted in hundreds of deaths and thousands of cases of serious illness. Farmers had received grain treated with mercurial fungicides for use as seed. Instead, they used it to make bread or fed it to livestock which they subsequently ate. Similar outbreaks occurred later in Pakistan and Guatemala. Canada banned the sale of mercurial seed dressings in 1970.

Government regulation of mercury

The cases of mercury poisoning did much to raise awareness of the risks of environmental mercury pollution. As a result, many countries created legislation and began monitoring waters in efforts to control mercury levels in the environment. In Canada, federal and provincial governments share jurisdiction over monitoring and controlling the use, production and emission of mercury.

Risks of mercury pollution uncertain

Unfortunately, it is difficult to assess the risks of exposure to small doses of environmental mercury pollution. There are presently no known biochemical early-warning signals that can be used to reveal small cellular changes occurring in the early stages of chronic mercury exposure. However, it seems clear that the people most likely to be affected by mercury pollution are those who rely heavily on fish and seafood in their diets, since that is the principal exposure route.

Indigenous people most at risk in Canada

This suggests that aboriginal people are probably at most risk in Canada. They rely heavily on fish both as a food and a source of income. As well, either through necessity or by choice, aboriginal people living on reserves often drink water directly from lakes and rivers. Both federal and provincial governments have carried out surveys of mercury levels in the blood groups of people in areas of high risk. Many residents of the Canadian North were found to have unacceptably high mercury levels in their bodies. For them, advice against eating contaminated fish threaten their livelihood and their way of life.

Food, way of life threatened

For example, mercury contamination in northwestern Quebec caused the Matagami fisheries to close, putting 200 people out of work. Similarly, commercial fishing was banned in the heavily contaminated English-Wabigoon River System in the 1970s, penalizing aboriginal people further through income loss. It was only in 1983 that the people affected were given any assurances of compensation. There is no way to put a price tag on the erosion of a way of life. For those communities that rely on local fish and wildlife for food, contamination of their fish deprives people of a meaningful part of their lives.

Arsenic

Low levels of arsenic needed for good health

An average person weighing approximately 70 kilograms carries about ten mg of arsenic in his body. The liver continually extracts this semi-metal from the blood, converts it to a less toxic form and then passes it out in urine. But we continually replenish this natural arsenic burden in our bodies because everything we eat contains some traces of it. Most organisms can tolerate minute doses of arsenic with no adverse health effects. In fact, small doses of this element seem to be beneficial, even essential, to life. Why it is necessary is still uncertain, but arsenic appears to stimulate growth and the production of haemoglobin in red blood cells.

Higher levels are poisonous

Ingestion of larger amounts of arsenic results in poisoning. This occurs mainly in industry and agriculture among workers who handle materials containing arsenic as an impurity. In these cases of acute arsenic poisoning, one of the symptoms is a painful paralysis of the lower limbs which resembles the polyneuritis (nerve inflammation) of the chronic alcoholic. Other symptoms include a scabby ulceration of the skin and a cafe au lait discolouration of the palms of the hands and soles of the feet.

Arsenic linked to cancer

The effects of chronic exposure to environmental levels of arsenic are essentially unknown and undetectable until the dose is high enough to elicit a serious symptom. It appears that the major threat associated with exposure to low doses of arsenic is the risk of developing cancer. Epidemiological studies have found very strong associations between cancer and exposure to arsenic. For example, a study of Quebec smelter workers who were occupationally exposed to arsenic, lead and cadmium reported cancer rates five to ten times higher than expected.

Drinking water can be major arsenic source

For the general population, the most important exposure route for arsenic is drinking water, especially groundwater. Some researchers estimate that the lifetime risk of developing cancer from drinking water containing as little as two mg/L of arsenic is one in 1000. If that is true, about 250,000 people around the world are suffering from cancers caused by chronic arsenic poisoning. Elevated levels of arsenic in drinking water are often derived from natural sources, but in some cases, mining activities have led to significant arsenic pollution of local waters.

Cadmium

Occupational exposure to cadmium

Cadmium and solutions of cadmium compounds are toxic, so that workers in the numerous occupations involving production or use of cadmium can expose themselves to dangerous levels of cadmium-containing fumes. Since 1962, Japan has recorded over 230 cases of degenerative bone disease attributable to cadmium poisoning from mine tailings. The Japanese name for the disease is "itai-itai" (ouch-ouch) disease because of the severe pain it causes in the joints.

Cadmium damages kidneys

The greatest hazard associated with exposure to cadmium pollution is kidney disease. Once again, there seem to be no definitive symptoms of chronic cadmium poisoning until the kidney damage is quite severe. Researchers estimate that between 250,000 and 500,000 people around the world may be affected by such cadmium poisoning. Cadmium is also implicated in cardiovascular disorders and cancers of the connective tissue, lung and liver. However, scientists are divided over the question of carcinogenicity. A study in England found no excessive cancer rates among inhabitants of a village where garden soil cadmium levels were nearly 200 times the average.

Greatest exposure is from food

Cadmium ingested in food is the largest exposure route for most people. Consequently, those most at risk are people living in areas where the soil is highly polluted. At present, the highest risks of environmental cadmium poisoning exist in Japan and central Europe.

Soil in Japan is contaminated with cadmium

Japanese soils are extensively contaminated with cadmium. As a result, researchers believe that the levels of cadmium ingested and stored in the kidneys leave residents of contaminated areas with very little margin of safety. Elderly women who have had several children are at the most risk and there are many women in this category who currently suffer serious kidney disorders.

Many soils in central Europe are also contaminated with cadmium. Denmark is one example where health officials are very concerned about the cadmium burdens accumulating in soils from atmospheric fallout and fertilizer use. Presently, the average intake of cadmium per day per person is 0.032 mg, but this is expected to double in the next century because of the increasing soil pollution. Concern about the health implications of this situation has led to various controls on the use of cadmium in products, the establishment of recycling programs for nickel-cadmium batteries and improved cadmium removal systems for incinerators and power-plant flue gases.

Concern about cadmium in Europe

URANIUM AND THE NUCLEAR FUEL CYCLE

Uranium is not usually discussed in surveys of metal pollution, although it is technically a metal. But while other metals are valued for such properties as malleability, strength or electrical conductivity, the greatest use of uranium by far is as a starting material for nuclear power and nuclear armaments. It is a very dense metal; because of this, uranium is also used for items such as balance weights in aircraft.

Uranium is also a metal

Mining and Refining

Uranium is a relatively common element in the earth's crust. Granite, for instance, which makes up about 60% of the Earth's crust, has an average of four ppm of uranium. Naturally occurring traces of uranium are found almost everywhere - in rocks and soils, rivers and oceans, as well as in food and human tissue. Because the uranium present in the environment is naturally radioactive, it contributes to the natural background radiation we experience every day.

Uranium is quite common in the earth

Like other metals, uranium is mined from ores. The largest potential for occupational and environmental pollution exists during the early stages of uranium production. Uranium ore is mined from either underground shafts or open pits. Explosives are used to break up the rock, which is then crushed further to facilitate transportation out of the mine to the milling site. In the mill, the uranium ore is ground into a fine powder and treated with chemicals such as sulphuric acid to dissolve the uranium from the powder. The uranium contained in this liquid is precipitated out as a uranium concentrate commonly known as "yellowcake".

Mining and milling of uranium ore

Very little uranium, relative to coal, is required for generating. However, the concentration of uranium is low even in commercially exploitable ores (from 0.1 to 10%), so that a considerable amount of rock must be moved. There is a greater than average potential hazard for lung cancer in uranium mines as the concentration of radon in such geological formations is greater than normal. There is thus a need for good ventilation in the mines to prevent a buildup of radon and radon daughters which could be inhaled by miners (see Chapter III).

Radon levels are high in uranium mines

Mining wastes stored in tailing ponds

The storage of waste tailings is probably the largest route for environmental pollution in the uranium fuel cycle. After the uranium has been mined and milled, large quantities of waste materials - both solid and liquid - are left behind. They occupy more volume than the rock removed during mining. These tailings are thus usually kept in tailing ponds on the mine site. The intention is to create an impermeable pond in which the solids and liquids will separate. The water is then either reused in the mill or treated to remove dangerous chemicals before being released back to natural waterways.

Water pollution from leaks in tailing ponds

Accidents have happened where tailing ponds have leaked or overflowed - contaminating the surrounding area with toxic chemicals and mildly radioactive substances. The Serpent River in northern Ontario was contaminated in the 1960s when acids seeped from the tailings into the river. The pollution seriously damaged the fish population and an aboriginal community downstream was affected.

CANDU reactors use natural uranium

Uranium refining and fuel fabrication involve the usual industrial risks. The process involves conversion of the yellowcake to uranium trioxide (UO_3). In order to power the Canadian natural uranium CANDU reactors, the trioxide is converted to the dioxide (UO_2). The dioxide is then formed into pellets used to fuel the CANDU nuclear plants developed in Canada. This accounts for the use of about one fifth of the uranium produced in Canada.

Most reactors need enriched uranium

About 99.3% of natural uranium is composed of the 238 isotope of uranium, with 0.7% of the 235 isotope. Uranium 235 is somewhat lighter than Uranium 238. The chemical properties of the two isotopes are the same, but they have different radioactive properties (see Appendix B). Although CANDU reactors can use natural uranium, most nuclear reactors in the world need uranium enriched to about 3.0% of the 235 isotope. To achieve this, the trioxide is converted to uranium hexafluoride (UF_6), the form required for isotope enrichment. Canada produces the hexafluoride for export, as there is no enrichment facility in Canada. Seven countries operate enrichment facilities. It is in these facilities that there is a potential for manufacture of weapons grade material.

Exposure to low level radiation

There is a risk from chronic exposure to low levels of radiation in the mining, milling and refining of uranium. The health effects of such chronic exposure are very uncertain, as the levels of radiation are often even lower than natural radioactive levels of material used to encase wastes. The health effects of exposure to higher doses of radioactivity are much more certain, but such high levels do not occur during the uranium production cycle. They have the potential to arise from the operation of nuclear reactors and the wastes generated by them.

Energy Generation

Growth in use of nuclear electricity

There has been a phenomenal growth in generation of electricity using nuclear energy in the last three decades. As of 1987, there were 418 nuclear generating plants operating in the world, with a total capacity of 308,000 megawatt (MW), and another 130 under construction with a capacity of 118,000 MW. Thirty-one countries had nuclear plants operating or under construction. The eight major users of nuclear power are shown in the following table.

Country	Operating Plants	Power (1000 MW)	% of Total Electricity
USA	109	100	15
France	49	47	70
USSR	57	34	10
Japan	37	28	30
W. Germany	21	20	30
Canada	19	13	15
UK	38	13	19
Sweden	12	10	40

Nuclear electricity generation

Electricity generation by nuclear and coal powered stations both involve the use of a thermal reactor. The essential difference between the two is the source of heat for the production of the steam driving the turbines. In coal reactors it is the combustion of coal to provide heat, carbon dioxide, water vapour and waste products. In nuclear reactors it is the fission (breaking up of nuclei of atoms) to form heat, radiation (see Appendix B), and new nuclear species. Many of the nuclear species produced are radioactive, for example iodine-131, cesium-137, strontium-90 and plutonium-239.

Nuclear fission produces heat, new elements

The fuel for heavy water (deuterium oxide) moderated reactors such as the Canadian CANDU can be natural uranium in the form of uranium dioxide, composed of over 99.3% of the 238 isotope. Natural uranium reactors (eg CANDU) do not require uranium enrichment but they do require a plant for the manufacture of heavy water (deuterium oxide). The greatest risk in CANDU reactor operation is most likely the effects of the possible release into the atmosphere of the hydrogen sulphide used in heavy water plants.

CANDU reactors need heavy water

Reactors in other countries use ordinary (light) rather than heavy water; they must use uranium enriched in the 235 isotope from 0.7% to 3.0%. It is also possible to use thorium or plutonium as nuclear fuels, rather than uranium. The plutonium must be "bred" in uranium by irradiation with fast neutrons. Six such "fast breeder reactors" (FBR) are now operating in three countries (France, USSR, Britain). Since the core of an FBR is compact, then a coolant with high heat removal capacity must be used; usually this is liquid sodium.

Plutonium is fuel for breeder reactors

The hazards of operation of a reactor arise primarily from radioactive releases due to accidents, both small scale and major ones like that at Chernobyl. In such accidents, radioactive pollutants can be spread into the environment.

Threats from nuclear accidents

There have been six accidents in nuclear power stations that have been so serious that part of the core of the reactor was damaged. The first actually occurred in Canada in 1952 when the NRX reactor had part of its core destroyed. An accident at Windscale in the UK in 1957 caused considerable radioactive contamination of the environment, but there is little evidence that other accidents, including the much-publicized one in 1979 at Three Mile Island

Serious accidents have occurred

in the US, resulted in large scale damage to human health or the environment. By far the most catastrophic accident was the one at Chernobyl in 1986, both in terms of deaths and the radiation released.

Chernobyl the most serious accident

The Chernobyl accident caused two immediate deaths when the explosion occurred. Of the 500 people with acute radiation sickness, 29 died within two months. The vast majority (465) of these patients were however at home and doing well one year later. It is however not these immediate acute effects that cause the greatest concern about the Chernobyl disaster.

Radioactive pollutants widespread

A large amount of radioactive material was released in the accident, much more than that released by nuclear weapons such as the bomb exploded over Hiroshima. Although the immediate environment was most heavily contaminated, there was significant fall-out of radioactive material as far away as northern Scandinavia.

Iodine-131 causes cancer of thyroid

The three radioactive isotopes that are of most concern in this contamination are iodine - 131, cesium - 137 and strontium - 90. Iodine - 131 has a short half-life of only about eight days, so that its hazards have become negligible in some weeks. Radioactive iodine is hazardous as it can be taken up by the thyroid gland in people, and can subsequently give rise to cancer of the thyroid.

Cesium and strontium more persistent

Cesium - 137 and strontium - 90 have half-lives of decades (see Appendix B). Thus contamination from these isotopes will persist for decades. These metals can be taken up by plants and animals, either through the plants or directly from dust which has settled out on the plants. Strontium is chemically similar to calcium and cesium to potassium, and both calcium and potassium are common elements in our bodies. The radioactive strontium and cesium can thus similarly be distributed throughout our bodies if ingested.

Many evacuated from Chernobyl area

There were 135,000 people evacuated from the most contaminated area within a month of the Chernobyl accident. More have since been evacuated. In contaminated areas not yet evacuated, people are forbidden to eat locally grown food and children are not allowed to play outside except on land which has been decontaminated.

Health effects from Chernobyl controversial

There are conflicting claims about the health effects arising from Chernobyl in the first five years. There is a claim that 7,000 of those who took part in the cleanup have died from exposure to radiation. This claim is countered by the observation that 600,000 people took part in the decontamination work, and that over five years, the number of deaths expected from a population of that size and age distribution would be expected to be about 7,000; the exposure to Chernobyl radiation may not have been responsible for any extra deaths. An international study of the health effects of Chernobyl was released in May 1991. The report stated that aside from signs of anxiety and stress, there was no statistically significant difference in the health of people living in areas contaminated by Chernobyl fall-out as compared to similar non-contaminated towns. Scientists and politicians from the regions affected argued vehemently against such a conclusion.

Uncertainty about long-term effects

The long-term effects of the accident are difficult to determine because of incomplete knowledge about the effects of low-level radiation, as well as the amount of radiation people received. Worldwide, it is estimated that during the next 50 years, between 1,000 to 500,000 (more likely between 5,000 and 75,000) people will die prematurely because of cancer. There may be as many as 700 children born with severe mental retardation due to radiation exposure in the foetal stage. Genetic abnormalities may occur in 10,000 people; these can be passed on to future generations.

Additional deaths may be hard to detect

It will be difficult to detect the expected increase in cancer deaths due to Chernobyl, even among the most seriously exposed, the evacuees. The collective radiation dose for the evacuated population has been estimated to be 16,000 sievert (Sv) (see Appendix B). Over the next 70 years, there would be about 160 extra cancer deaths (one for each 100 Sv according to Appendix B) in addition to the 14,000 expected normally. This translates into an expected increase of less than two per cent in the death rate from cancer among the Chernobyl evacuees, those who were most seriously affected.

Increased cancer in the future

For the 75 million people in the vicinity, the additional dose due to iodine-131 was estimated to be 290,000 Sv, raising the cancer rate in the next decade by about 0.05%. Iodine-131 has a short half-life of 8.5 days, but the cancer it may cause can have a latency period of over a decade. The effect of cesium-137 is expected to be larger. The 2.1 million Sv to which people are likely to become exposed is expected to increase the incidence of cancer deaths by 0.5% in the next 50 years.

Monitoring of exposed children

In the nine months following Chernobyl, there were about 300 births in the immediately affected area. In the case of teratogenic effects, the time at which the exposure took place is important, as well as the total dose. There has not been any increase noted in the number of cases of mental retardation, but the children are being carefully monitored for a longer period. No ill effects have been noted for foetal doses of radiation below 100 mSv.

Reactions in countries affected

Countries affected by the radioactive fall-out from Chernobyl reacted with varying speeds and in a variety of ways to the radioactive contamination of their environment. The most common actions taken by European countries were: advice to wash fresh vegetables before eating, advice not to drink rainwater, removal of milk-producing cattle from grazing, and import restrictions on food from some countries. Grazing cattle efficiently collect radionuclei deposited on grass over a wide area, and milk is commonly consumed by children, who are more susceptible to damage by radiation. Some foods, such as reindeer meat, still had considerable contamination from the Chernobyl fall-out over a year later.

Further consequences of Chernobyl

This discussion of the effects of the Chernobyl accident has centred around the effects on human health from radiation. There are other damaging consequences; to animals and the ecosystem as a whole; to the peace of mind and the traditional lifestyles of the people evacuated or exposed; to the restriction of children's activities; to the economic costs of the cleanup.

Spent Fuel and Nuclear Waste

International restrictions on technology

The technology for uranium enrichment is of critical importance and is therefore restricted by international agreement. It is only when uranium is sufficiently enriched in the 235 isotope that it can be used to manufacture nuclear explosives. Plutonium obtained from processing spent fuel or from "breeding" can also be used. The potential for nuclear power to be misused for military purposes arises when enrichment or reprocessing plants are available.

Spent fuel is highly radioactive

The fuel in a conventional nuclear reactor becomes highly radioactive, and is removed when only a few per cent of its potential fission energy has been used up. This is because nuclei build up in the fuel elements which prevent nuclear reactions from occurring effectively. Such "spent fuel" is normally stored in large water tanks until the radiation from shortlived radioisotopes has diminished. It can then be reprocessed to form new fuels, or it can be considered as a "waste" material and sent to disposal. Because of the high radioactivity of the spent fuel, reprocessing can be a hazardous operation. There are three commercial reprocessing plants in operation, two in France and one in Britain. The record of the British Windscale plant in polluting the environment through discharges is much worse than those of the French reprocessing plants.

Dismantling of old plants hazardous

There are hazards involved in the eventual disposal of radioactive wastes. This includes the dismantling of nuclear power plants which have become too old or obsolete to maintain in a safe condition. One of the hazards encountered in dismantling the Windscale plant was the presence of large amounts of asbestos (see Chapter III). The presence of radioisotopes in the metal parts still presents the greatest hazard.

No final answer to disposal of nuclear wastes

The eventual disposal of nuclear wastes is still under discussion. The hazards are long-term, as the wastes will remain dangerously radioactive for a period of time spanning generations. There are fears that future generations will not take adequate precautions with this waste. The radioactive materials could also leave their containers and spread through streams or groundwater to cause radioactive pollution over a wide area. A public review started in Canada in 1990 of a concept for disposal of high-level reactor wastes in deep rock formations.

CONCLUDING REMARKS

Metals are present everywhere

Metals are present throughout the earth - in rocks, soils and sediments. Some of the metals we consider hazardous are among the more common elements in the earth's crust. These metals are, for the most part, locked firmly in place in some reasonably stable form. They nevertheless do enter the air and water in more mobile forms due to natural processes such as weathering or volcanic eruptions. This situation is actually desirable, since many metals are needed for life on the planet. Some elements which are very toxic in larger amounts are essential to our physical well-being in small amounts.

In this industrial age, we have learned to exploit the properties of metals to make our lives healthier, more comfortable, even more exciting. Our dependence on metals results in widespread mining of ores, accompanied by smelting and other extraction, purification and transformation processes. As a result of such industrial activities, we extract metals from their stable locations and release them into the environment. We live in a time when human activities are more significant than natural processes in mobilizing nearly every metal locked in the earth's crust, leading to environmental pollution by numerous metals.

Mining releases metals from stable forms

Unfortunately, such pollution is effectively permanent. The atoms of a metal cannot be altered by chemical or biological action - they have an infinite life-time in the environment. This sets metals apart from even the most persistent pollutants such as PCBs or DDT which, even though very long-lived, do eventually breakdown in the environment.

Metals do not biodegrade

This chapter focuses on five metals which are the cause of major concern. In all five cases, the possible chronic effects from exposure to environmental pollution is the main hazard. But such effects are difficult to find and difficult to substantiate, which can make it difficult to obtain the necessary support for legislation and regulation.

Chronic effects are the main concern

There is evidence that many people have carried more than a "natural" burden of lead in their bodies since as long ago as Roman times. This may not have had any detrimental effect on our health, but we cannot be completely sure. Low-level lead in the environment is so globally pervasive that it may not be possible to find a control group any more with natural lead levels.

Everyone may have an un-natural burden

In the past three decades, we have become more aware and more knowledgeable about the chronic effects of metals like lead, mercury, cadmium and arsenic in the environment. In many countries appropriate legislation has been passed to lower the emissions of these metals. There are nevertheless surprises in store. Few people anticipated that one of the consequences of the James Bay hydroelectric project would be mercury contamination in the fish that the Cree relied on for a major part of their diet.

Awareness of risks has led to regulations

In the case of the uranium fuel cycle, environmental pollution can occur at several stages: from mine tailings, from reactor releases, from waste depositories. Nuclear energy remains a very controversial issue because of the nature of the risk involved in operating a nuclear reactor. That is, the consequences of an accident can be very high, but the probability of an accident happening is very low. In Canada, there are several lines of defence built into nuclear power reactors (see "AND" gates, Chapter VIII). Only in the extremely unlikely case where several well maintained, thoroughly tested safety systems fail simultaneously would there be a serious accident.

Nuclear energy remains controversial

In the case of mining residues and waste disposal, there is a greater probability of a release of contaminated water, dust or other material. However, the consequences of such releases are very small since the material is not very

Smaller risks from low-level radioactivity

radioactive. Only if highly radioactive material is involved is there a more serious consequence to human health and the environment, and in that case more precautions are required to lower the probability of a release.

Public involved in waste management

Currently, the greatest potential hazard to the environment from our nuclear energy program would seem to be from the tailings of mining operations. The plans for nuclear waste management are being developed very slowly with major public input in order to reduce the actual risk, and also the anxiety, that could be caused by a waste depository.

Recommended Reading

"Blighted Land, Blighted Lives", J. Webb. *New Scientist*, pp 20-21, 15 June, 1991.

"The Legacy of Chernobyl", M. Bojcun. *New Scientist*, pp 30-35, 20 April, 1991.

"Reassessing Nuclear Power: the Fallout from Chernobyl", C. Flavin. Worldwatch Paper 75, March 1987.

CHAPTER VIII
RISK ASSESSMENT OF CHEMICALS

Types of hazards from chemicals

We are concerned about chemicals in the environment for several reasons. A variety of direct health hazards are associated with chemicals: they can cause direct and immediate illness, injury or death; they can cause illness or death after long-term, low-level exposure; they can cause birth defects (teratogenicity) or changes in genetic material (mutagenicity) which are passed on to future generations.

Hazards to the environment

Chemicals that are used in our society can also cause damage to the environment and to structures built by humans. Such damage can be immediate, for example the destruction caused by explosions or chemical spills. It can also be of a long-term nature, through degradation of the quality of the environment or irreversible losses to ecosystems as in the case of species extinction. Human welfare can also be affected if an impaired ecosystem can no longer meet our needs, and if resource sustainability is lost.

Risk measures the seriousness of a hazard

The magnitude of hazards is expressed in the form of "risk". Risk is usually considered as the probability of the occurrence of an event multiplied by the magnitude of some negative consequence that results. A specific example would be the probability of a release of a chemical in a loading platform times the damage (human, material or environmental) caused by such an event. The risk of another type of chemical hazard may be the probability of exposure to a substance times the consequence of such an exposure. An example of the risk of this type of hazard would be the concentration of radon a person is exposed to in a house multiplied by the expected health effect of this exposure.

Risks are accompanied by benefits

Rational people, governments and companies knowingly undertake activities involving risk only when there is some benefit involved. We take the risk of crossing a busy street only if we derive some benefit, such as getting home, from this action. In some cases, risk is regulated without regard for the benefit involved. This has been the case in the United States for chemical food additives. Under the "Delaney amendment", no substance was allowed to be added to food if it had been found to be carcinogenic in animal tests. The benefits of the substance were not considered. Some substances, which are on the "generally regarded as safe" (GRAS) list, were exempted.

Money is a common measure

In the cases where risk and benefit are weighed against each other, it is necessary to have a common measuring unit. Thus the magnitudes of risks (or costs) and benefits must be estimated; this is usually done in terms of dollars.

Hazards are not immediately identified

There are numerous problems involved in estimating risk. One is that the hazard may not even be identified. It took decades before any negative effects of chlorofluorocarbons and PCBs were recognized. Only in the last decade has the seriousness of the health effects of second-hand smoke been accepted. The presence of mercury in rivers and lakes was not perceived as a hazard until the 1950s. Even though we are now more alert to the possibility of hidden hazards and carry out more thorough assessments, there may be hazards from chemicals of which we are not yet even aware.

Radioactivity was believed to be safe at first

Another example of ignorance of a hazard is the case of radioactivity, which is now accepted as a proven hazard. However, it was not believed to pose any risk until many years after its discovery. Wilhelm Roentgen discovered x-rays in 1895, and they were used for various applications soon after, with little concern for safety. Over three decades later, in 1927, it was found that an exceptionally large proportion of the marriages of radiologists were childless, presumably due to the mutagenic effect of the exposures to x-rays they experienced in their work. The dose-effect relationship for radiation is still not well established. The International Atomic Energy Agency altered it quite dramatically as late as 1990 (see Appendix B).

PERCEPTION OF RISK

Individuals perceive risks differently

The issues discussed in this book are often controversial because people have many differing perceptions of the risks involved. The above examples illustrate cases where no risk is perceived at all. In the long term, this often causes great difficulty. For example, because PCBs were not considered to have any potential hazards, they were allowed to build up to a considerable extent in the environment. Similarly, chloroform was used for decades in medicines because it was not believed to be hazardous.

Differences in Values

Difficulties in quantifying risk

Estimating the magnitude of a risk is difficult because the negative effects can be very different in nature: death, illness, property damage, ecological change, etc. Some people will consider a potential change in the environment to be much more important than the illness of other people. In order to reach a common estimate for the magnitude of risk, usually expressed in dollars, it is necessary to consider controversial issues such as the value of life (or life-years), or the value of a biological species.

Personal values affect risk estimation

Even though people have the same information available, their individual risk estimates of a chemical in the environment can differ considerably, as our quantification of the negative consequences can differ widely due to differences in our personal values. Many such values can be expressed through two types of "discounting" which we carry out, either consciously or subconsciously.

Estimation affected by time involved

"Time discounting" expresses the importance we place on the future relative to the present. People with a high time discount rate consider the consequences of an event to be much smaller if it occurs in the future rather than right away. Such people would regard the potential harm twenty years in the future of increased UV radiation resulting from CFCs released today to be very small (see

Chapter II). Those with a zero discount rate would consider the future harm to be as important as if there were no delay. Thus, if we are convinced that CFCs will eventually lead to serious negative consequences, and if we have a very low time discount rate, then we would estimate the risk of continued release of CFCs to be great.

Proximity of effect is of importance

"Geographical discounting" is similar to time discounting. If we care little about people far from us, then we would estimate the magnitude of a risk to be low if it affects people removed from our own community. The problems of chemicals in our sewage would seem less serious if we have a high discount rate, since communities down-stream would be affected rather than our own.

Characteristics of the Hazard

Other factors affecting risk perception

Our anxiety about a hazard and our acceptance of it depends on our perception of the risk involved. The risk we perceive is essentially our individual implicit estimation of the risk. It can differ from an actual risk estimate as obtained through an objective, statistical quantification even in the absence of differences in the values we place on negative consequences, as discussed above. A number of factors can influence our perception:

Voluntary or involuntary?

* Is the risk voluntary? Studies have shown that people are willing to accept much greater risks if they are voluntary, such as smoking, than if they are involuntary, such as sitting in a room with second-hand smoke.

Are benefits and risks shared fairly?

* Are the risks and benefits shared fairly? Communities are reluctant to accept a waste disposal site or an incinerator in their neighbourhood if the waste comes from another community. They may be willing to accept the risks only if they are compensated for it, that is, if they obtain some benefits.

Do we have control over the hazard?

* How much control do we have over the hazard? We are more likely to accept a risk when we are in control (driving a car) than if someone else is in control (flying in an airplane). We are very anxious about the possible hazards in our municipal drinking water, over which we have little direct control, while we readily accept the risks involved in drinking our favourite beverage.

Can the hazard be easily eliminated?

* Can the hazard be easily eliminated? It is relatively easy to decrease CFC emissions; decreasing carbon dioxide is much more challenging. CFCs are manufactured only by a few companies, their use is not always essential, for example the CFCs in deodorant sprays, and reasonable alternatives are possible. However, carbon dioxide emissions come from very many sources, most of which are energy-related. It is essential to have a source of energy, but alternatives such as nuclear energy are perceived by many to carry great risks and alternatives such as solar and/or conservation are limited.

Are we familiar with the hazard?

* How familiar is the hazard? The risks of familiar hazards are not perceived to be as great as those of unfamiliar ones. Deaths from car accidents are relatively common, yet the perceived risk of death from auto accidents is less

than the actual risk. The media has a large influence in determining the familiarity of hazards. High consequence, low probability risks generally cause more anxiety in society than warranted.

Availability of Information

Accurate information is important

From a societal point of view, it would be best if the perceived risk or anxiety were reasonably similar to the actual risks. A desire to reduce our anxiety is what motivates us in our personal actions. Governments and industries are also likely to respond to the anxiety of the public, even though such actions may not be the most efficient in reducing the actual risk. If we are to be reasonably efficient in using our resources to reduce actual risks, then it is important to have a well-informed society where the magnitudes of various risks are perceived reasonably accurately.

The media play a crucial role

The function of the media is very important in this. There have been cases where the public became extremely anxious about some chemicals which, in fact, presented a very low risk. The concern about formaldehyde in urea-formaldehyde foam insulation (UFFI) was, in retrospect, seldom justified (see Chapter III). In the last decade, PCBs have caused much anxiety, but again in retrospect, there is little evidence that they have caused any significant human health effects in North America (see Chapter V).

No controls unless hazard is recognized

In cases where hazards are not yet identified there is no perceived risk even if there is an actual risk. These situations of blissful ignorance are certainly not desirable. We obviously can not take any steps to lower unknown risks. Moreover, once we identify the hazard, such as the effect of sidestream tobacco smoke on people or of CFCs on stratospheric ozone, it may be difficult to reduce the risk of the hazard. There may be a large amount of the chemical already in the environment. An industry dependent on the chemical may have become established, and it would have an interest in maintaining itself. It is moreover difficult to re-educate the public about the safety of a familiar chemical.

Anxiety is undesirable

Unnecessary anxiety is undesirable for yet another reason. Anxiety causes unhappiness and stress, and why should we be unhappy without any real reason? Moreover, stress itself can have negative health consequences.

ESTIMATION OF RISK

Acute and chronic exposures possible

People can be at direct risk from a chemical in two different ways. "Acute" exposure is when there is a sudden contact with a large dose of the chemical. This can be through ingestion, through inhalation or through the skin. Often, it is due to some accident. For this reason, analysis of accidents will be discussed later in this chapter. Such acute exposure can cause injury, burns, illness or death. People can also be harmed by long-term exposure to low levels of the chemical; this is called "chronic" exposure. It can give rise to illnesses such as cancer, and can thus cause great anxiety.

Acute Exposure

Acute hazard data from animal tests

The risks to humans from acute exposure for a few chemicals can be derived from historical data. For most chemicals, however, the risk is estimated from experiments carried out on animals, usually rats. A common test is to give a group of animals a large single dose of the chemical, usually expressed as milligrams of chemical per kilogram body weight. The object is to determine the dose at which half (50%) of the animals die. This is called the "lethal dose 50%" or LD_{50} level. Tests are also carried out to obtain "lethal concentration 50%" or LC_{50} levels. This can be for fish or animals with the chemical present in water or air respectively. After a time period (typically chosen to be four hours), the percentage of dead fish or animals is noted.

The LD_{50} levels for oral doses are given in the table below for some chemicals and some animals. The list is chosen to illustrate some points made following the table.

Acute toxicities of selected chemicals

Chemical	Animal	LD_{50} (mg/kg)
Arsenic trioxide	rat	15.
	human	1.4
Dioxin	rat	0.02
(2,3,7,8-TCDD)	hamster	5.
	guinea pig	0.0005
Caffeine	rat	192.
Nicotine	rat	50.
2,4-D	rat	370.

Human data available for arsenic

This table shows data on humans for arsenic, as it has been involved in a sufficient number of homicides and suicides to supply historical evidence. Rats can tolerate a dose that is 10 times as great as humans. Such variation of toxicity with species is more evident in the case of dioxin. Hamsters can tolerate considerably more dioxin than rats, but dioxin is extremely toxic to guinea pigs.

Some common chemicals are quite toxic

Caffeine (in coffee) and nicotine (in tobacco) are chemicals which many of us ingest voluntarily every day. They are both significantly more toxic to rats than the herbicide 2,4-D, but it is not the acute effects of 2,4-D that are causing concern to people (see Chapter VI).

Chronic Effects

Chronic effects cause much concern

Acute effects are important, but the concern about most chemicals in the environment is raised by chronic exposure that may lead to cancer. There are considerable difficulties in estimating such risks where the probability of exposure is high but the consequence is low due to the low level of exposure. In contrast, risks from acute exposure have a low probability, but a high consequence.

Chemicals can have threshold limit values

There are two approaches to estimating the risks of chronic exposure. In the case of some illnesses, it is possible to define a "no observed effect level" (NOEL) from animal studies. This usually occurs where the illness is due to some physiological process such as inhibition of an enzyme by the chemical. In such cases, an "acceptable daily intake" (ADI) can be derived for substances to be ingested, or a "threshold limit value" (TLV) can be set for substances in air. In Canada, this is usually done by dividing the NOEL level observed in animal studies by a safety factor of 100. This allows for variations in the toxic effects to humans compared to animals and for variations between individuals.

There may be no threshold for carcinogens

The concept of a NOEL may not be valid for processes such as the formation of cancers, teratogenesis and mutagenesis. Such effects may be due to a single random event where a molecule of the chemical tested causes a mutation of the DNA in a cell, initiating a chain of events that leads to cancer. The chemical structure of a molecule often suggests how it or its metabolites will act in living cells.

Carcinogenicity correlates with mutagenicity

The correlation often found between cancer and mutation makes it possible to use "mutagenicity assays" to identify possible carcinogenic chemicals. In such an assay, the frequency of mutations caused by the chemical in a strain of bacteria is determined in a relatively rapid, inexpensive experiment. Such a test detects mutagenicity in about 90% of the carcinogenic substances examined.

Animal tests for chronic effects

Animal tests are a better indicator of carcinogenicity to humans than the tests described above which use primitive organisms such as bacteria. However, the animal tests are expensive and take a long time. Typically, groups of 50 rats are exposed to at least two different doses of the chemical being tested over a period of about two years. The rats are then killed and examined for malignant and benign tumours. A comparison is made with a control group, where the rats received a zero dose of the chemical.

Utility of tests at maximum tolerable dose

The highest dose the animals are given is the "maximum tolerable dose" (MTD), so that the effects of the chemical are more readily observable. Such doses can be ridiculously high as compared to those that humans can expect to be exposed to. Indeed, it is possible that a dose as high as the MTD is such a serious insult to the animals' physiology and causes so much stress that tumours may be formed due to the stress itself rather than the chemical. Thus MTD experiments may, in cases, not be relevant to the situation in which human exposure is likely to occur. This issue is currently under intense debate.

Statistical significance of tests important

Even with high doses, the number of tumours may not be very large. In such cases, it can be difficult to decide whether the results are statistically significant. If one out of 50 rats in the control group developed a tumour, as compared to three of the 50 in the highest dose group, should the chemical be considered a carcinogen? There is a possibility that the three tumours were formed by chance, just as the one in the control group. A 95% significance level is generally accepted in the analysis of the results. There will thus be a small possibility that a chemical will be branded a carcinogen due to the chance formation of tumours, and vice versa.

Because of the high doses administered experimentally, the health effect of a chemical at the low doses actually encountered is difficult to estimate. This is done by extrapolation on a "dose-effect" curve. The result is very dependent on the shape assumed for the curve. The shape of such a curve depends on the mechanism with which cancer develops, and this is not fully known. Several models exist, among them the "threshold" model, which assumes a "no effect" level (curve a). The most common assumption used is that the dose-effect curve is linear (curve b).

Extrapolation of dose-effect curves

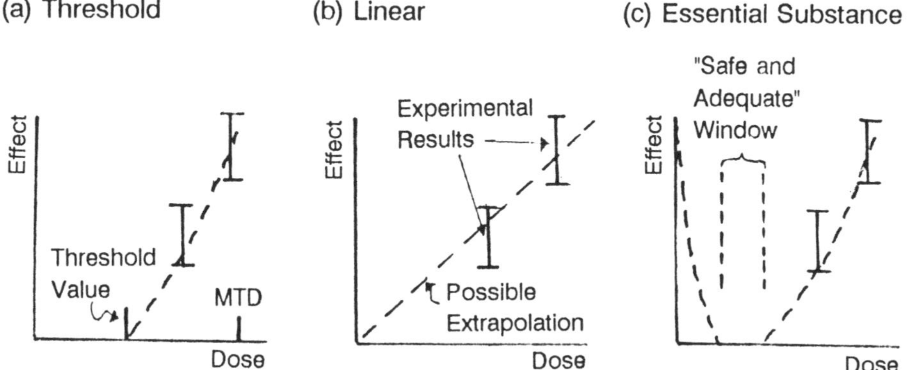

Three examples of dose-effect curves

Some substances which show toxic effects (arsenic, selenium) are also essential in small amounts to sustain life. Their dose-effect curve would be something like curve c. We all need traces of them to survive. Some essential vitamins, such as vitamin A, can also be toxic if taken in excessive amounts.

Some chemicals are essential for life

Tests are usually carried out for individual chemicals. The effects of chemicals can, however, be synergistic. For instance, the risk of exposure to asbestos by a person who is a smoker is greater than the sum of the risks of exposure to asbestos and of being a smoker taken separately. There is evidence that PCBs and DDT also act synergistically in insects.

Toxicity can have a synergistic effect

In using data from chronic animal studies for estimating risk in humans it is also assumed that our reaction to the chemical is the same as for the animals. Studies using both rats and mice show some differences as indicated in the following table. It gives the daily oral dose (in milligrams of chemical per kilogram body weight) required for a lifetime in order that half the animals develop tumours, the "toxic dose 50%" (TD_{50}).

Humans are assumed to react as rats

Chemical	Animal	TD_{50} (mg/kg)
Benzene	rat	157.
	mouse	53.
Chloroform	rat	119.
	mouse	90.
Formaldehyde	rat	1.5
	mouse	44.
Ethanol	rat	9110.
Aflatoxin	rat	0.003

Toxic doses for selected substances

Variability of data for rats and mice

Data on chronic effects such as those shown in the table are not available for humans. The variability between humans and animals can thus not be assessed. Rats and mice do not show much difference in their reaction to benzene and chloroform, but formaldehyde is far more carcinogenic to rats to than mice. Generally, rat data are about 85% correct in predicting carcinogens for mice.

Some natural substances are also toxic

Some substances formed in nature can be potent carcinogens. Aflatoxin is present in moulds which can grow on foods such as peanuts. It is extremely effective in giving rise to tumours in rats. Ethanol and formaldehyde are also natural substances. Ethanol is present in trace amounts in many natural foods since fermentation is a natural process. Formaldehyde is present in all our bodies since it is part of a basic human biochemical cycle (see also Chapter III).

Little human evidence for cancer causes

Sometimes the type of evidence presented above is the only suggestion of carcinogenicity in humans. There is good evidence that ingestion of aflatoxins by humans has led to increased cancer incidence. There is also some evidence from occupational exposures that benzene exposure causes leukaemia in humans. However, no clear evidence is available that groups of people who have had high exposures to chloroform (see Chapter V) or formaldehyde have a greater incidence of cancer.

Data suggest behaviour modification

If we base our behaviour on the results of animal TD_{50} data such as those in the table, we come up with surprising results. We would be foolish to worry about the traces of chloroform present in drinking water if we continue to use even very small amounts of alcoholic beverages. A bottle of beer would be about 10,000 times as dangerous as a glass of average tap water.

Animal studies can save human lives

The testing of chemicals with animals as described above is an attempt to prevent harm to humans in the first place. This has been successful, since for eight of the 54 known human carcinogens, the evidence was first obtained from animal experiments.

Epidemiological Evidence

Control group needed in epidemiology

The ultimate evidence of the carcinogenic nature of a substance is human epidemiological evidence. In an epidemiological study, a search is made for correlations of particular diseases with habits, genetics, climate, diet or environmental factors. In order to ensure that the suspected substance is the cause of the disease, a control group is required. This should be a group which resembles the group under study in all aspects except exposure to the substance studied.

Study of risk from asbestos convincing

It was an epidemiological study that showed convincingly that exposure to asbestos caused cancer of the lung as well as a form of cancer specific to asbestos called mesothelioma (see Chapter III). This required analysis of the cause of death in a group of 17,800 men who had strikingly high occupational exposure to asbestos dust from 1967 to 1977. There were about 4.6 times the number of deaths from lung cancer (486 in all) in this group than would be expected in a group of that size not exposed to asbestos fibres.

Epidemiology linked cancer to smoking

One of the figures below shows another example of epidemiology using historical data. Lung cancer death rates in American males have increased dramatically, while stomach cancer rates have fallen. The rise in the incidence of lung cancer is believed to be due to the increased incidence of smoking. It is delayed by more than a decade because many types of cancer have long latency periods. The other figure shows a correlation between meat consumption and colon cancer in various countries.

Examples of epidemiological correlations

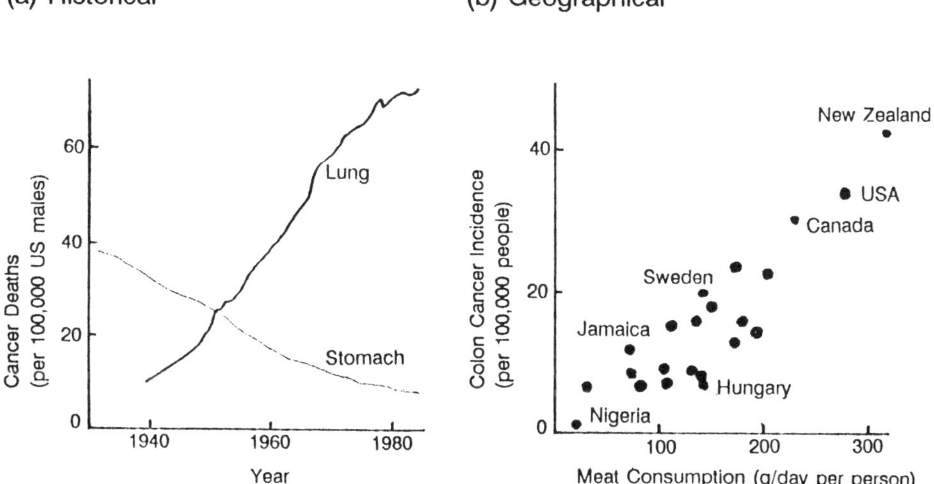

Correlation does not mean cause

Correlations may point to a cause and effect, but there are difficulties with such interpretations. For instance, there is a good correlation between the numbers of births and of storks in Germany, but this obviously does not mean that storks bring babies. Increased industrialization is more likely to be the common cause for lower birth rates and fewer nesting sites for storks.

Difficult to determine cause, effect

The baby/stork example is an obvious fallacy in determining cause and effect in a correlation. There are, however, many cases where correlations have not been easy to resolve. The inverse correlation between lead levels and academic performance in working class children could be because higher lead levels result in poorer brain function. But it could also be that children who play on streets spend less time in intellectually stimulating home environments. Thus, they would have higher lead exposure and levels; but the lower academic performance might be due to a less intellectually stimulating home environment.

Problems in collection of data

Collection of data is often difficult and uneven in time or between regions. Problems of data acquisition become even more difficult when latency periods are present. The occupational exposure history of people involved in a study has to be estimated many years in the past.

Sample choice can influence results

The choice of population sample is also crucial. The larger the sample, the more likely that the result can be statistically significant. However, the significance of a result is also affected by the level of exposure; it can be lowered if a smaller sample of highly exposed people (eg petrochemical workers) is "diluted" with people who had low exposures (eg secretaries and managers at the plants).

Epidemiology has pitfalls

Because of such complexities, epidemiological evidence is not as easy to obtain and interpret as it would seem at first glance.

ESTIMATING RISK FROM ACCIDENTS

Accident risks from historical evidence

For some types of accidents, it is possible to evaluate the magnitude of the risk from historical evidence. Accidents which have occurred during the transport of chlorine, and the effects of various concentrations of chlorine gas on humans are known and recorded. Thus the risk to humans of transporting chlorine can be evaluated, at least in principle.

Problems with use of historical data

There are however problems with the above evaluation. The historical data on chlorine releases through transport accidents may have little bearing on the probability of accidents in the future. After an accident, steps are usually taken to try to prevent a repeat occurrence. New safety regulations, new training programs, and better emergency response measures are often implemented. Some of the historical data may also have little relevance if taken from situations different from that of concern. The frequencies and types of railway accidents in countries such as Mexico and Canada can be very different due to a variety of factors such as regulations, technology, training and climate.

There are no data for some accidents

There are many situations where there have been very few, perhaps even no, accidents. Such a situation is a very desirable, but the risk involved in operating an accident-free chemical plant or shipping chemicals in such a seemingly safe way must nevertheless be estimated. The industry, the regulators and the public must have an idea of the degree of safety in the design and operation of such systems. Even if the probability of an accident is very small, the consequence can be enormous; thus the risk can be considerable.

Modelling is used when data is lacking

In the absence of historical data, the risk analysis must proceed by modelling. Usually data are available on the probability of failure of the components in a chemical plant, such as pumps. Such data can be obtained from testing or from day-to-day operations. The probability of operator errors in various tasks can also be obtained from history or experiment.

Risk assessment using fault-trees

An analysis of a system can be done by building a "fault-tree". Each failure causing hazards can be traced back through the branches and twigs of the "tree" to the failures of components and operators for which data are available. Examination of such fault-trees can show where the placement of more safety devices would be most effective. One limitation of fault-tree analysis is that we can never be sure that we have accounted for all possible failure modes.

Two possible causes for an error

To illustrate how a fault-tree works, consider the factors involved in assessing the probability of a chain of events leading to a train crash where a chlorine tank car is ruptured and people are exposed to chlorine gas. Suppose a train mistakenly passes a signal. First, examine the possible reasons for this error: it could be because of a failure in the signal (past data might show there is one chance in 10,000 of this event); or because of driver error (also one chance in 10,000). Thus, the total probability of a train passing a signal would be one chance in 5,000 (obtained by adding the two separate probabilities).

Two faults occurring at the same time

Just passing a signal would probably not cause an accident, unless an obstacle such as another train is present on the tracks. We can assume the probability of that other train being present is one in ten. Now we see that, a little further into the fault-tree, the probability of wrongly passing a signal and having another train present is one in 50,000 (obtained by multiplying these two probabilities).

Several chances to avoid accident

Even this level of analysis is not sufficient, since the train may have time to stop. Unless, of course, the brakes fail - something that may happen once every 10,000 times they are applied. Even once we have come to the branch in the fault-tree where there is a collision, there may or may not be a rupture of a chlorine tank. Finally, even if there is a rupture of a tank, there still may not be any exposure of people to chlorine if the crash is in an unpopulated area.

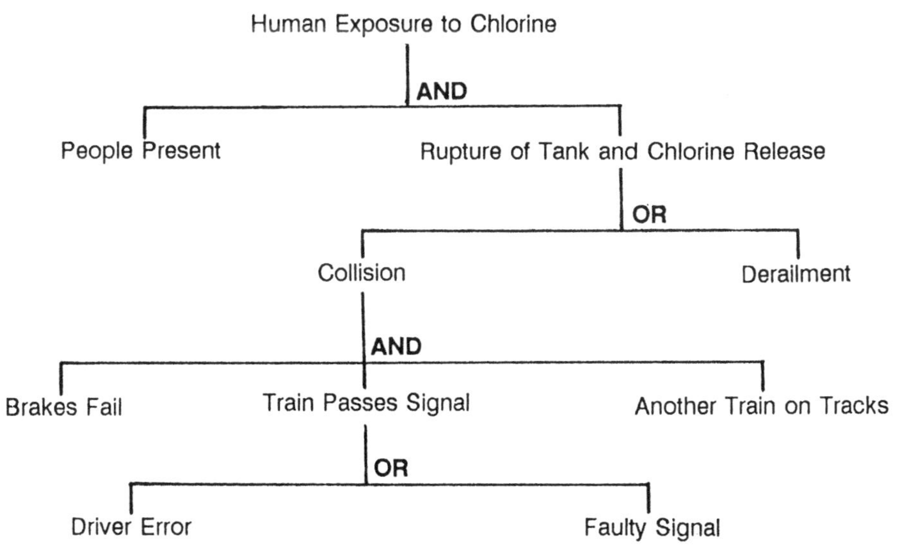

Fault-tree of harm from chlorine release during transport

All possibilities must be considered

All possible chains of events leading to an exposure of people to chlorine must be included in a fault-tree. The overall probabilities of different magnitudes of exposure must then be worked out by appropriate additions (in "or" gates) or multiplications (in "and" gates) as illustrated above. In actual fact, most ruptures of tank-cars are due to derailment. Thus the "branch" of the fault-tree that we discussed above is not the one needing most attention.

Probability-consequence curves change

The results of such analyses are often presented on "probability-consequence" diagrams such as the one shown below. Note that plane crashes with a consequence of 10 fatalities occur between one and 10 times a year; thus historical evidence for this phenomenon is available. Meaningful historical evidence is essentially unavailable for events which occur less than once a decade. The technology and regulations change significantly for events such as plane crashes and chlorine shipments. Thus, the curves shown are to a large extent hypothetical, obtained from modelling.

Examples of probability-consequence curves

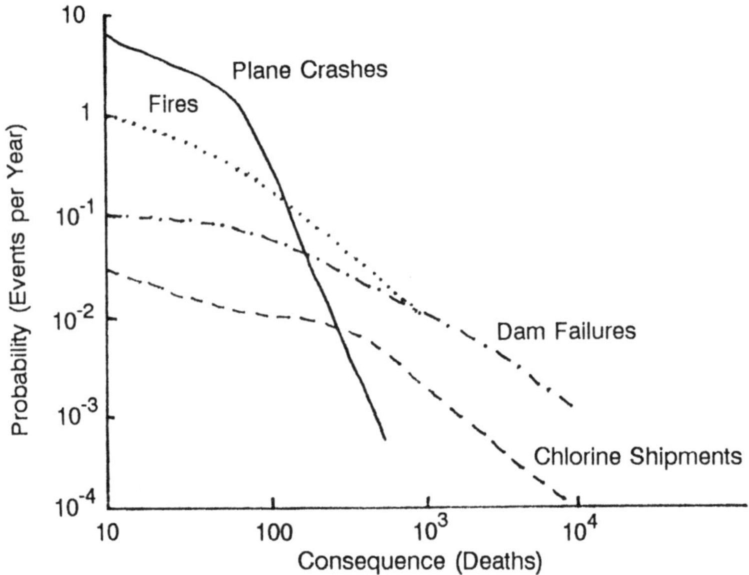

Operation and maintenance of safety systems

The above type of analysis must be done in the design of an acceptably safe plant, transport system, or the like. The designers must include sufficient safety devices and back-up systems to make a plant acceptably safe. In order to achieve an acceptably low risk, the design of the actual plant must have integrated with it the design of the operating and maintenance procedures, as well as the training of the people who will run the plant. The tragic accident at Bhopal where over 3,000 people died may have been due to improper maintenance and operation of safety systems.

RISKS TO ECOLOGY

Estimation of ecological risks difficult

Estimating the risks to ecological systems is perhaps even more difficult to carry out than estimating the risks to human health as discussed above. How can we estimate the probability of the various scenarios of the future global climate? How can we estimate the negative consequences of such changes to the ecology? How can we put a value on the loss of biological diversity, deforestation, and other ecological changes?

Computer models used as tools

Estimations of these changes and their risks are also done by modelling. Models of the global climate are very complicated, and even with the availability of supercomputers many effects must be left out. Only now are climate models sufficiently advanced so that some biological effects can be included in them.

The effects of global climatic change can also be estimated by considering historical data. There have been considerable changes in the global climate in geological time, and evidence of the changes is preserved in sediments and ice. Data on more recent changes are also available from trees and meteorological records.

Some historical data are available

Some of the risks involved in global climatic change are very uncertain. Yet their magnitude is so great that it would be folly not to give them serious consideration. Other risks are very certain, but of a much smaller overall magnitude, such as the risk of illness from exposure to chlorine. It is difficult to decide how much of our resources we should use to lower a risk that is uncertain but great versus one which is certain but small.

Uncertainty of risk estimations can vary greatly

When we consider the evaluation of risks in the far future and in distant places, the notion of discounting becomes especially important. If we are to make decisions guided by the relative magnitudes of different risks, then we have to compare their values at the same point in time. If we discount with regard to time, then what is likely to happen in the distant future becomes negligibly important. With a time discount rate of 10% per year, something likely to take place eight years from now is only half as important as if it has the same likelihood of occurring in one year's time.

Time discounting important in risk estimation

The problem of geographic discounting is also difficult to handle. In a case such as CFCs, actions taken in one country may have benefits or risks in another. The development of refrigeration in a country such as China or India may be affected by regulations on CFCs made by, and primarily for the benefit of, the industrialized world. Emissions from industrial plants and coal burning electricity generating stations in Eastern Europe affect the ecology of Scandinavia. Who should pay for the clean-up?

Who should bear the cost of risk reduction?

Such discounting problems are difficult to solve. Each individual, each culture, has their own idea of how important the consideration of the future should be, and how much consideration we should give our neighbour, our fellow humans in a distant land, or even creatures other than *homo sapiens*. There is no universally correct answer. The benefit-cost analyses carried out by economists may at least give a framework for considering such problems.

How important are other species?

MANAGEMENT OF RISK

How should risk be managed? In nearly all cases, resources are required to reduce risk, and it is more and more difficult to reduce a risk the smaller it becomes. How large should a fire department be? How strong should a bridge be? How small should the exposure to a chemical be? How many safety systems should there be in a chemical plant? The safer we wish to be (that is, the smaller the risk) the greater the price we must be prepared to pay.

How safe is "safe enough"?

We can never achieve zero risk

One thing is certain: there is no such thing as a zero risk. There is always a background risk from natural hazards such as lightning or meteorites. This risk is of the order of one death in a million per year or between one in 10,000 and one in 100,000 per lifetime. Discussion about the level of "acceptable risk" or "effectively zero risk" in California seems to indicate that it will be defined there to between one in 10,000 and one in a million in a lifetime.

Data on cancer deaths

Data from the US show that approximately 25% of the population eventually dies of cancer, or 250,000 people out of one million. If exposure to a level of a chemical raises the lifetime risk of death from cancer by one in a million, then 250,001 people (25.0001%) will die of cancer in a group of one million.

Increase in risk can be difficult to detect

The large background rate of cancer deaths makes this increase seem insignificant. It has been estimated that if the US Environmental Protection Agency (EPA) were to regulate all carcinogens in air, soil and water to insignificant levels, then the decrease in the cancer incidence would be between 0.25% and 1.3% of the annual total cancer deaths. The accuracy of death statistics is probably too poor to note such a small change.

In time, small effects add up

A risk of one in a million per year (25 people per year in Canada) seems small compared to deaths from traffic accidents or homicides. The overall consequences in large populations can nevertheless be significant over a number of years. Thus, if there is no significant benefit from a chemical, and if its concentration in the environment can be lowered without enormous cost, then the attempt to reduce its presence should be made.

We can detect ever smaller amounts

No honest, competent chemist can guarantee that there is zero concentration of any chemical in the environment. The more sensitive analytical methods become, the more capable we become in detecting even the slightest traces of chemicals in environmental samples. If two decades ago no dioxin could be detected in a lake, then it was only because we did not have techniques sensitive enough to detect it. Today, much smaller amounts can be detected; next year even smaller amounts will be detectable. As the years go by, the number of potentially hazardous chemicals detected in our air and water will increase - simply due to more sensitive analytical instrumentation.

Accident risks remain

In the same way, no honest, competent engineer can say that there is zero probability of a transport accident or an accident in a chemical plant. Even with the most sophisticated and careful modelling, it is impossible to design a completely accident-proof plant or transport system. A very small risk always remains.

Benefit-cost analysis for managing risk

Benefit-cost analysis can be used in deciding an acceptable level of risk reduction. However, as shown earlier, the quantification of risk in terms of dollars is full of assumptions and value judgements. When we can quantify risk, then a good management criterion may be to reduce the risk to where the cost of additional risk reduction is greater than the benefit gained from the reduced risk.

Risk may be acceptable if benefit great

It may prove wise in some cases to take great risks if the potential benefits are even greater. Patients with AIDS may thus be willing to undergo a treatment which cures their disease even if the treatment doubles their chance of getting cancer. Risk-benefit considerations are of central importance in drug treatments and surgery.

Placing a price on a human life

In many cases it may be necessary to place a price on human life and suffering. Assigning such values may seem immoral, but refusal to do this can not only lead to inefficient decision-making, but also to a net loss of human lives. We can end up using hundreds of millions of dollars introducing new technology or regulations which may save one life per year. The same resources used for different purposes may result in the savings of far more lives or in increasing the quality of life for thousands of people. There is a great difference in the estimation of the price of an identified life as compared to an unidentified (statistical) life. The discussion here deals with the price of a statistical life.

Reducing risk can itself involve risk

Spending a hundred million dollars to save one life is counter-productive. There are many hazardous activities involved in the manufacture of a safety system: mining, transport, welding, to name a few. German statistics report that the expenditure of about thirty million dollars results in activities leading to one fatality.

Possibility of counter-productive behaviour

Risks involved in the activities and materials used to reduce risk cannot be neglected. If the cost of reducing the risk by one fatality is greater than some tens of millions of dollars, then it is likely to be counter productive to reduce the risk any further. Using this criterion, some technologies and materials may turn out to be so risky that they should not be allowed. In other cases, we may be insisting on so much protection that we are counter productive in maximizing life and health.

Anxiety can lead to ill health

There is even risk in warning messages. Stress and psychosomatic effects can have greater ill effects than the actual chemical or the potential accident. In retrospect, there was probably more harm caused by poor information after the accident in Seveso, Italy, than by the chemicals released into the environment (see Chapter V).

The public commonly mistrusts experts

The management of risk has often been done by experts, excluding the public and the press. This used to be the case to a large extent in Canada some decades ago. It is now more common to have public involvement through Royal Commissions, public hearings and input from special interest groups. The trend seems to be to distrust experts in industry and government. As indicated earlier, the experts have in the past failed to identify the hazards of numerous chemicals which they introduced, for instance CFCs, PCBs, lead in gasoline and certain pesticides. The participation of new groups in risk assessment may improve this record. The assessment of risk can also be very subjective and the final decision is often dependent on the value system of the people involved. Experts in risk assessment have no mandate to decide on appropriate value systems for people.

COMPARISON OF RISKS

Utility of comparing risks

The previous discussion has indicated how difficult it is to compare risks. The difficulties are greatest when the risks are of differing nature. Nevertheless, some comparisons are useful as a guide to see which types of technologies or regulations should be developed for the most efficient use of our resources for lowering overall risk. Comparisons of the magnitudes of risks of the same type are more meaningful.

Environmental Protection Agency Rankings

EPA suggestions for risk priorities

The EPA in 1990 released a report setting priorities for risks; the actual magnitudes of risks were not quantified. A majority of the hazards they considered involved chemicals in the environment; these considerations are summarized here. The Canadian situation is not dramatically different from that in the US. The EPA divided the risks into two main categories: risks to human health, and risks to the natural ecology and human welfare.

Human health effects include cancer

In risks to human health, both the risks of cancer and non-cancer effects were considered by the EPA. Their ordering from highest risk to lowest in terms of chemicals in the environments is as follows:

Outdoor air pollution risks high

* Ambient air pollutants, including those from mobile and stationary sources (sulphur and nitrogen oxides, carbon monoxide, ozone and other components of smog). This category also includes metals such as lead and arsenic which reach us by various pathways involving the atmosphere.

Indoor air pollution next

* Indoor pollution, including radon and combustion products such as nitrogen oxides and tobacco smoke. Indoor exposure to hazardous chemicals in consumer products is also included, such as formaldehyde and solvents.

Water risks low

* Pollutants in drinking water, including lead and chloroform

Ranking of risks to ecology

In terms of risks to the natural ecology and human welfare, the EPA considers the following as high-risk problems: depletion of stratospheric ozone, and global climatic change through emission of greenhouse gases. As medium risk problems they list: pesticides, nutrients and toxic chemicals in surface waters, acid deposition, and airborne toxics. The following problems are considered to have relatively low risk: oil spills, ground water pollution, radionuclides, and acid run-off.

Time discounting affects ranking

The ranking of one of the above lists versus the other depends to a large extent on the rates of discounting discussed previously. With zero time and geographic discounting, the risks to the ecology and human welfare are likely to predominate over the immediate risks to human health.

Environmental ethic affects priorities

There is considerable debate about the relative value of individual humans versus the ecosystem. Some argue for a "human-centred" view, ie that the ecosystem is of importance primarily for the purpose of sustaining the welfare of humans. Others maintain that the ecosystem as a whole is what matters; individual species such as *homo sapiens* are not of crucial importance. Still others believe that we should not practise "specism", ie that all species capable of consciousness should have equal rights. Such different philosophical points of view make it challenging, to say the least, to reach consensus on major projects involving technology, society and the environment.

Comparison of the Risk of Death

The magnitude of risks can be expressed in different ways. There are advantages to each method of comparison, but pitfalls in all of them.

Several ways to compare risk

The first table gives the average risk of death in the US as expressed in probability per year. Some of these risks are well known from historical evidence. In other cases they are uncertain because they are based on animal test results.

Evidence for risk estimates differs

Examples of risk estimates for death

Cause	Risk	Uncertainty
Cigarette Smoking (one pack/day)	$3.6x10^{-3}$	Factor of 3
Motor vehicle accident (total)	$2.4x10^{-4}$	10%
Air Pollution (eastern USA)	$2x10^{-4}$	Factor of 20
Home Accident	$1.1x10^{-4}$	5%
Alcohol (light drinker)	$2x10^{-5}$	Factor of 10
Background Radiation (excl. radon)	$2x10^{-5}$	Factor of 3
Peanut Butter (4 tbs/day)	$8x10^{-6}$	Factor of 3
Water with EPA limit of chloroform	$6x10^{-7}$	Factor of 10

The risk of drinking water containing the EPA limit of chloroform is thus estimated to be from $6x10^{-6}$ to $6x10^{-8}$. The large uncertainty is due to the lack of human epidemiological evidence.

Estimate of risk from chloroform uncertain

The next table expresses risk in terms of the loss (gain) in life expectancy from various selected causes.

Risk expressed as change in life expectancy

Cause	Loss (Gain) in days
Smoking (one pack/day)	1600
Being a coal miner	1100
Motor Vehicle Accidents	200
Accidents in the home	95
Legal drug misuse (in US)	90
Falls	40
Poisoning	17
Installing smoke alarms in homes	(10)

Yet another way of comparing risks is to compare activities that increase a risk by an equal amount. The table lists activities which increase the probability of death in any year by one in a million. This data is again derived from US statistics and animal data.

Tabulation of activities with equal risk

Activities which give the same increase in risk

Activity	Cause
Smoking 1.4 cigarettes	Cancer, heart disease
Living 2 months with smoker	Cancer, heart disease
Drinking 1/2 litre wine	Cirrhosis of liver
Spending one hour in coal mine	Black lung disease
Travelling 500 km by car	Accident
Living 2 days in New York City	Air pollution
One chest x-ray	Cancer from radiation
Eating 100 charbroiled steaks	Cancer from PAHs
Drinking Miami water, one year	Cancer from chloroform
Eating 40 tbs of peanut butter	Cancer from aflatoxin

Risk comparison difficult, controversial

There are, as illustrated, many ways of comparing risk. In making comparisons, it is important to remember that some risks are better known than others. It may be possible to meaningfully determine the relative risks between two hazards of the same type, for example two air pollutants causing cancer. However, when the hazards are different in nature, even comparing risks becomes more dependent on the value system of the people involved and increasingly full of potential for controversy.

Recommended Reading

"Readings in Risk", T.S. Glickman and M. Gough, eds. Resources for the Future, Washington, DC, 1990.

"Reducing Risk: Setting Priorities and Strategies for Environmental Protection", United States Environmental Protection Agency, SAB-EC-90-021, September, 1990.

"Chemicals, Cancers, Causalities and Cautions", B.N. Ames. *Chemtech*, pp 590-598, October 1989.

"State of the Environment Report for Canada", Environment Canada. Ottawa, 1986.

"Risk Analysis in the Chemical Industry", Chemical Manufacturers Association. Government Institutes, Inc., Rockville, Maryland, 1985

APPENDIX

A : THE ELECTROMAGNETIC SPECTRUM

Light, x-rays, heat and radiowaves are all forms of electromagnetic radiation. They simply have different frequencies. The major source of such radiation is the sun, which sends out electromagnetic radiation in a vast range of frequencies. Fortunately, the upper atmosphere intercepts most of the radiation harmful to humans. Solar radiation is reflected and absorbed by various substances. When it is absorbed, the energy can be re-radiated, but at a lower frequency. Technological activities also produce large amounts of electromagnetic radiation.

The figure on the next page shows the complete electromagnetic spectrum in terms of both frequency and wavelength of the radiation. The visible range of radiation occupies only a very small part of this spectrum, flanked by the infrared on one side and the ultraviolet on the other. A summary of the various categories of radiation is given below, indicating their potential hazards to humans. They are in order of increasing frequency, ie decreasing wavelength. The unit used for frequency is the hertz (Hz), which is one cycle per second.

Extremely Low Frequency (ELF) Radiation

The major source of ELF radiation (range 0 to 300 Hz) is the 60 Hz field arising from high voltage transmission lines and electrical appliances. There are conflicting reports of the effects of ELF on the nervous system, blood chemistry, increase in cancer rates and increase in abnormal pregnancies. No definitive relationship between ELF exposure and harm to humans seems to have been established to date, but research into this potential hazard continues.

Radiofrequency and Microwaves

The range of radiofrequency is 100 thousand Hz (100 kHz) to 300 million Hz (300 MHz), corresponding to a range of wavelengths from three thousand metres (3 km) to one metre. Microwave radiation is defined to be between 300 MHz and 300,000 MHz corresponding in wavelength from one metre to one-thousandth of a metre (1 mm).

Broadcasting and telecommunications, microwave ovens and hospital equipment are the major sources of this radiation. Some unintentional emission can also be present from electrical appliances and lights.

No adverse effects have been noted in people who have received medical treatments with microwave radiation. Injuries have occurred to animals in experiments where they have been exposed to intense radiation as this causes increases in temperature. Effects to the nervous system of animals on long-term low-level exposure have also been noted, as well as genetic and teratogenic effects in animals and plants.

Optical Radiation

This includes infrared (IR), visible and ultraviolet (UV) radiation. The wavelength of IR is between one millimetre (1 mm) and 760 billionths of a metre or nanometres (760 nm). Visible radiation has wavelengths between 760 nm (red) and 380 nm (violet). UV is between 380 nm and one nm. UV-A has a range from 320 to 380 nm; it is not energetic enough to give us a suntan. The range of UV-B radiation is between 290 and 320 nm; it is energetic enough to cause tanning of the skin.

Only radiation with a wavelength less than 100 nm is energetic enough to cause ionization of biological material. The range 380 nm to 100 nm is non-ionizing UV, while from 100 nm to one nm is extreme, or vacuum UV, which is ionizing. Some references consider ionizing UV radiation to be a part of the x-ray spectrum.

Sunlight is the major source of radiation in this range. Essentially all the ionizing radiation is filtered out before it reaches the troposphere. Most of the UV-B is absorbed by ozone in the stratosphere, but some remains at sea level. Lamps, electrical heating, flames and lasers are technological sources of such radiation, but each source of this type emits a narrower range of radiation than the sun. Lasers are sources of very intense radiation at a specific wavelength.

Optical radiation in the non-ionizing range has been shown to cause a number of injuries to humans such as accelerated skin aging, sunburn, skin cancer, cataracts, and retinal injury. These hazards are present from both technological sources as well as from natural sources, ie solar radiation. The hazards of ionizing radiation are essentially the same as those of x-rays.

X-Rays and Gamma Rays

X-rays have wavelengths between 100 nm (10^{-7} m) and one-hundredth of a nanometre (10^{-2} nm or 10^{-11} m), while gamma rays have wavelengths smaller than one-tenth of a nanometre (10^{-1} nm or 10^{-10} m).

X-rays are produced for medical and industrial uses by bombarding a target with electrons; but they can also be produced unintentionally in some electrical appliances. Gamma rays are present in cosmic radiation and are emitted in the radioactive decay of nuclei, whether natural, eg in rocks, or in technological systems, eg nuclear reactors.

Both types of radiation have been shown to cause cancer, as well as teratogenic and genetic effects. The risks of exposure to this range of electromagnetic radiation are discussed further in Appendix B, since the same units of biological damage (rem or sievert) are used as for nuclear radiation.

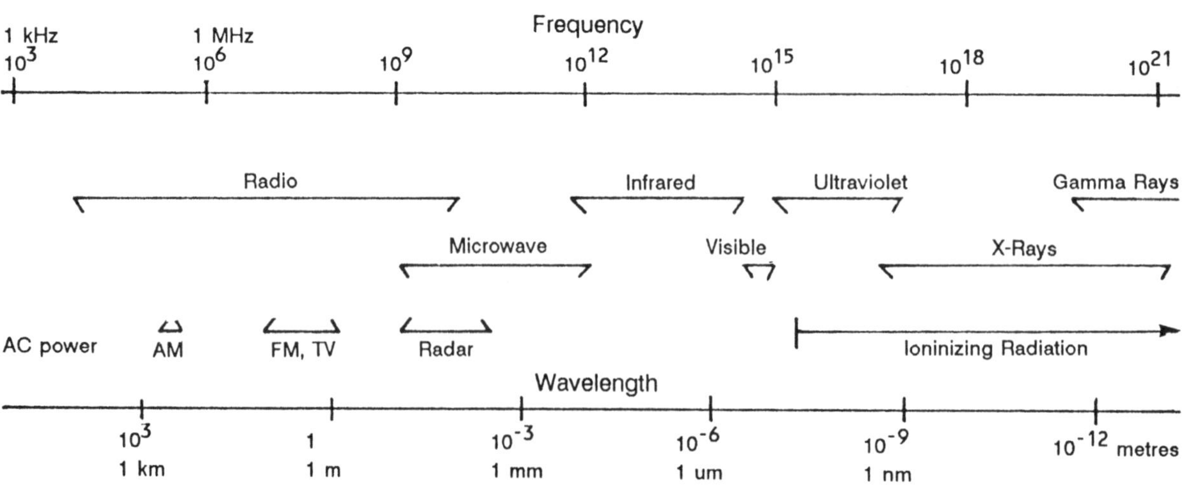

The Electromagnetic Spectrum

B : NUCLEAR RADIATION

Gamma rays, a form of electromagnetic radiation, are emitted directly from reactions of atomic nuclei. There are two other common forms of nuclear radiation: alpha, consisting of particles composed of two neutrons and two protons; and beta, consisting of electrons. They are often referred to as alpha and beta "particles" rather than "radiation".

Alpha particles do not penetrate far into materials; a thin sheet of paper or the dead layer of cells on the surface of the skin is usually sufficient to stop them. But they can cause a great deal of damage if they are emitted inside the body from inhaled or ingested materials. Beta particles can penetrate much further into human tissue; they can pass through a hand. Gamma radiation is even more penetrating and can pass right through the body, just as x-rays.

Alpha, beta, gamma and x-ray radiation all have sufficient energy to break chemical bonds in biological materials, causing them to break up into fragments which are electrically neutral (radicals) or charged (ions). This can damage biological cells irreversibly, causing death or initiating cancer.

Half-Lives

The half-life of a radioactive isotope is essentially a measure of the amount of time it takes for it to lose half its radioactivity. An isotope with a short half-life of 8.5 days such as iodine-131 (see Table) is highly radioactivity, but only over a short time. After 8.5 days, the radioactivity of a sample of iodine-131 will be half that at the start; after 17 days it will be one-quarter, etc. This is of importance in considering the time-scale of an effect, eg the contamination of food by the Chernobyl accident.

The most common isotope of natural uranium, 238, has a very long half-life of 4,500 million years. This means that the radioactivity of a sample of natural uranium will diminish so slowly with time that in a human time-scale it will essentially be constant. There will also be so little radioactivity emitted that it will present essentially negligible health hazards.

Isotope	Half-life
Polonium-218	3.0 days
Radon-222	3.8 days
Iodine-131	8.5 days
Cobalt-60	5.3 years
Strontium	25 years
Cesium-137	33 years
Carbon-14	5700 years
Plutonium-239	24,100 years
Uranium-235	710 million years
Potassium-40	1,400 million years
Uranium-238	4,500 million years

Half-lives of Isotopes

Units for Ionizing Radiation

There are two sets of units for describing ionizing radiation and its effect on humans. The Standard International (SI) system (becquerel-gray-sievert) should now be used, but the old system (curie-rad-rem) is also given below.

The becquerel (Bq) refers to the radioactivity of a sample; it corresponds to one disintegration per second of any nucleus. The gray (Gy) is the SI unit of absorbed dose of energy from radioactivity. One Gy corresponds to one joule of energy from the ionizing radiation absorbed per kilogram of tissue. In the old system, energy absorption was expressed in rad; one Gy equals 100 rad. The type of radiation, ie whether alpha, beta, gamma or x-ray, is not considered in this measurement of dose.

A dose of alpha radiation can cause considerably more biological damage than the same dose (in Gy) of gamma radiation. A unit has thus been introduced for expressing the potential biological damage due to radiation, or the effective dose. The SI unit is the sievert (Sv), while the former unit is the rem, which is short for "roentgen equivalent man" (the roentgen was the unit used for measuring the dose of x-rays).

Gamma and beta rays are about equally damaging to biological tissue as x-rays; they thus have a "quality factor" of unity. A dose of one Gy of gamma radiation will thus have a potential damage of one Sv, (one rad has one rem). Alpha particles have a quality factor of 20, as they are about 20 times more damaging per dose as x-rays. One Gy of alpha particles thus corresponds to 20 Sv. Note that one Sv equals 100 rem, and that one millisievert (mSv) equals 0.1 rem.

Exposure to Radiation

The table lists the major sources of radiation. The total effective dose from natural sources is of the order of two mSv per year. This varies considerably with location. At higher altitudes, the contribution of cosmic radiation will be greater. The radioactivity of rock and air varies considerably with geology.

The contribution to the effective dose from medical treatment is estimated at 0.30 mSv/yr, but this obviously varies with the number and type of x-ray examinations and treatments. Fallout from atmospheric tests of nuclear weapons, primarily in the 1950s, is believed to contribute 0.010 to 0.024 mSv/yr. The contribution from air travel is variable; for a Paris to New York trip it is 0.04 mSv with subsonic aircraft. Even though cosmic radiation levels are greater at higher altitudes where supersonics fly, the overall effective dose for a supersonic trip is less because of a shorter time for exposure. The contribution from nuclear power has been estimated to be 0.0001 to 0.002 mSv/yr, depending on location.

The effective dose from technological sources is thus very variable, and is primarily determined by exposure to medical x-rays.

Source (major isotopes)	Effective Dose (mSv per year)	
Cosmic radiation	0.30	
Rocks, soil (radon, potassium, uranium)	0.45	
Food forming our bodies (potassium, carbon)	0.30	
Atmosphere (radon)	0.80	
Total natural		1.85
Medical	0.30	
Nuclear weapons' test fallout	0.02	
Air travel	0.01	
Consumer products (watches, TV)	0.01	
Nuclear power	0.002	
Total technological		0.34

Sources of Radiation Exposure

The International Commission for Radiological Protection altered their recommendations for the annual effective dose limit for members of the public in 1990. They lowered their recommended annual dose limit from five mSv/yr (about two and one-half times the natural background) to one mSv/yr. The recommended limit for the dose to radiation workers is now 50 mSv/yr for one year; such occupational exposure is considered to be voluntary and financially compensated. The total dose should not exceed 100 mSv over a five year period.

The risks of cancer due to low levels of radiation are difficult to determine. The rule of thumb is that the lifetime risk of fatal cancer per sievert of exposure is one in 100. This implies that natural background radiation is responsible for about 1.4% of all deaths. An exposure of five mSv/yr thus gives an extra risk of 50×10^{-6} or 50 in a million per year of fatal cancer. The risk of genetic disease is estimated to be between two and four per thousand per sievert exposure.

C : CHEMICAL FORMULAS

Carbons and hydrogens are not shown in most of the complex structures. Each carbon has 4 bonds; those not shown are hydrogens. Thus

denotes cyclohexane

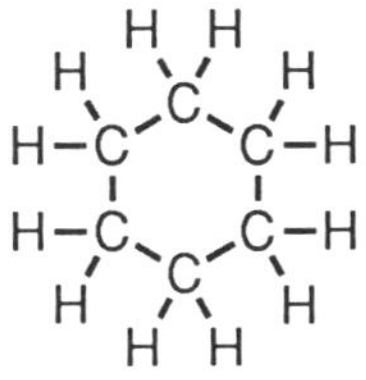

H H
H C H
H–C C–H
H–C C–H
H C H
H H

and

denotes benzene

H
H C H
C C
C C
H C H
H

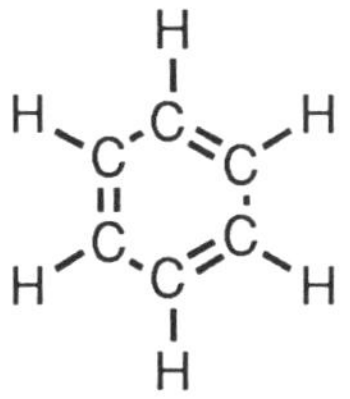

Alachlor
2-chloro-2',6'-diethyl-N-methoxy-methylacetanilide

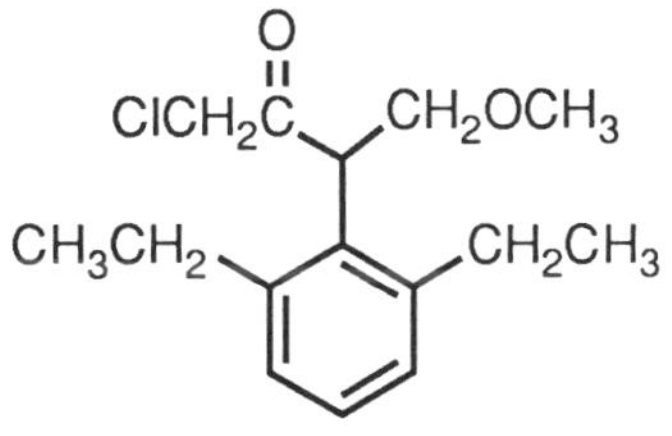

O
‖
$ClCH_2C$ CH_2OCH_3
CH_3CH_2 CH_2CH_3

Aldicarb
2-methyl-2-(methylthio)propionaldehyde-O-methylcarbamoyl oximine

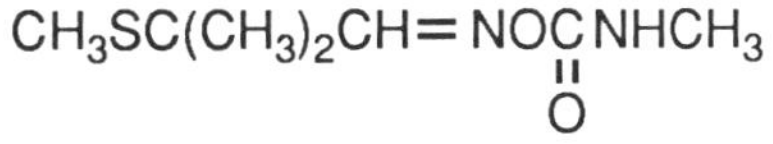

$CH_3SC(CH_3)_2CH{=}NOCNHCH_3$
‖
O

Asbestos
chrysotile, white asbestos

$3MgO{\bullet}2SiO_2{\bullet}2H_2O$

coridolite, blue asbestos

$Na_2O{\bullet}Fe_2O_3{\bullet}3FeO{\bullet}8SiO_2{\bullet}H_2O$

Atrazine
6-chloro-N-ethyl-N'-(1-methylethyl)-1,3,5-triazine-2,4-diamine

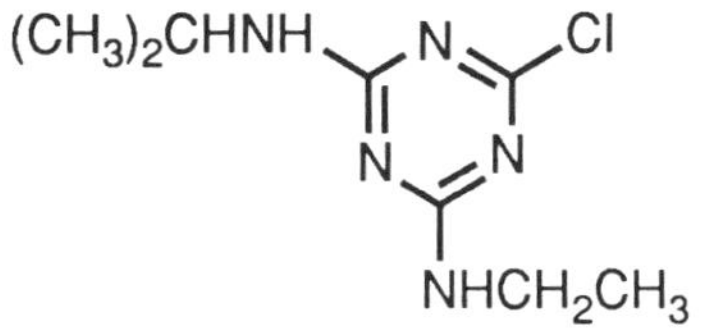

$(CH_3)_2CHNH$ N Cl
N N
$NHCH_2CH_3$

Borax
Sodium Borate

$Na_2B_4O_7$

CFCs (chlorofluorocarbons)
CFC-11, trichlorofluoromethane

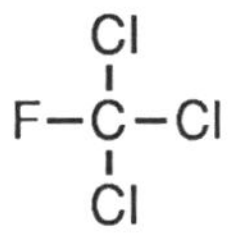

Cl
F–C–Cl
Cl

CFC-12; dichlorodifluoromethane

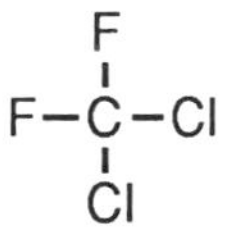

F
F–C–Cl
Cl

CFC-113; 1,1,2-trichloro-1,2,2-trifluoroethane

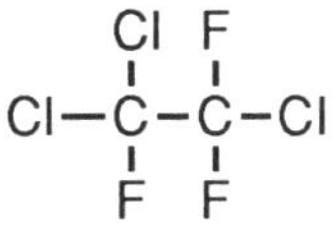

Cl F
Cl–C–C–Cl
F F

Chloropicrin
trichloronitromethane

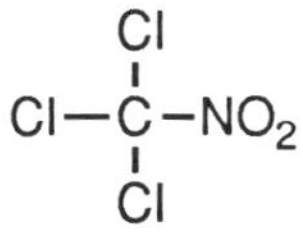

Cl
$Cl{-}C{-}NO_2$
Cl

DBCP
1,2-dibromo-3-chloropropane

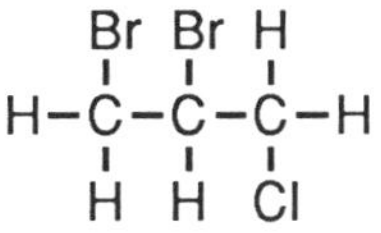

Br Br H
H–C–C–C–H
H H Cl

DDE
1,2dichloro-2,2-bis(*p*-chlorophenyl)-ethane

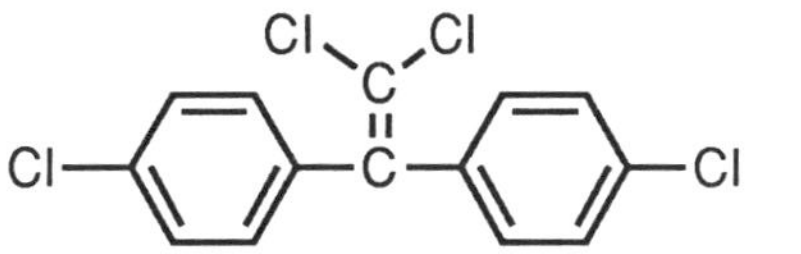

Cl Cl
C
‖
Cl– –C– –Cl

DDT
1,1'-(2,2,2-trichloroethyliline)bis-[4-chloro-benzene]

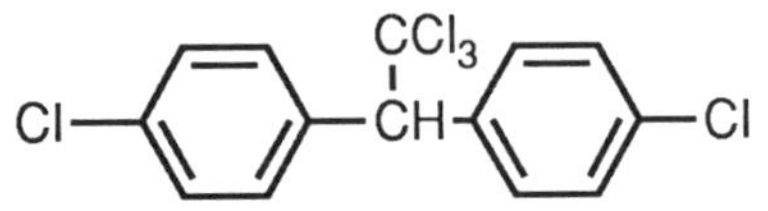

Dieldrin
3,4,5,6,9,9-hexachloro-1a,2,2a,6,6a,7,7a-octahydro-2,7:3,6-dimethanonaphth[2,3-b]-oxirene

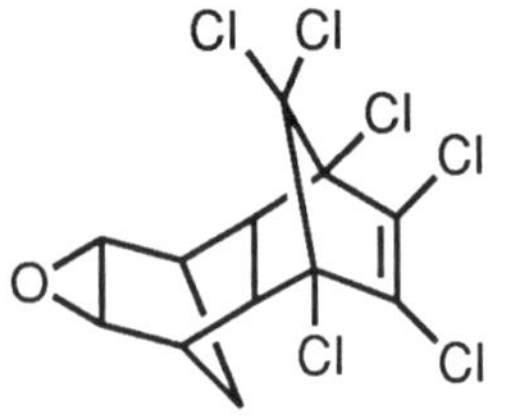

Diholoacetonitriles
dichloroacetonitrile

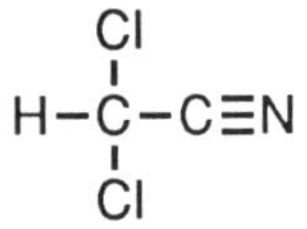

$$H-CCl_2-C\equiv N$$

DMN
dimethylnitrosamine

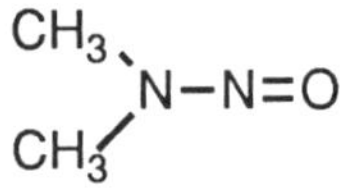

$$(CH_3)_2N-N=O$$

DMS
dimethylsulfide

$$H_3C-S-CH_3$$

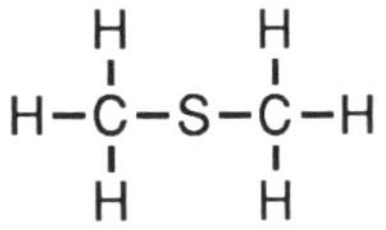

Fenitrothion
Phosphorothoic acid O,O-dimethyl-O-(3-methyl-4-nitrophenyl) ester

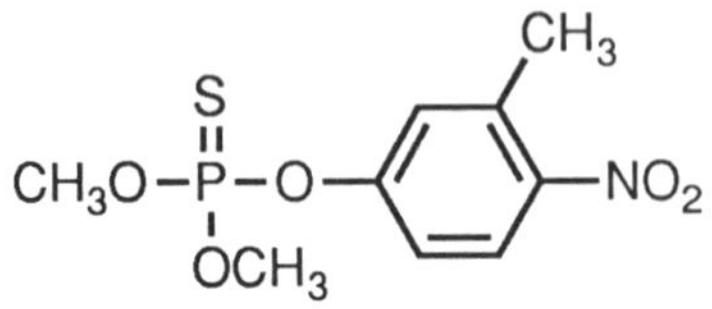

Formaldehyde

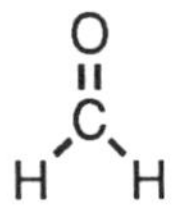

$$H_2C=O$$

Halons
difluorobromochloromethane

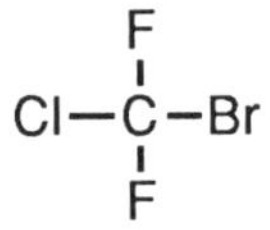

$$Cl-CF_2-Br$$

trifluorobromomethane

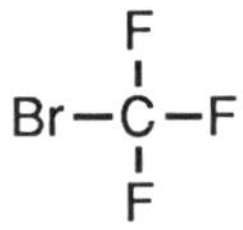

$$Br-CF_3$$

HCFCs (hydrochlorofluorocarbons)
1,2-dichloro-1,2,2-trifluoroethane

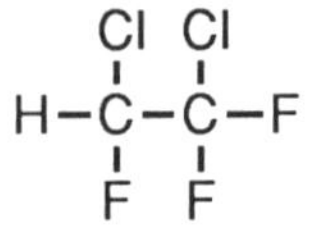

$$H-CClF-CClF_2$$

HFCs (hydrofluorocarbons)
1,2,2,2-tetrafluoroethane

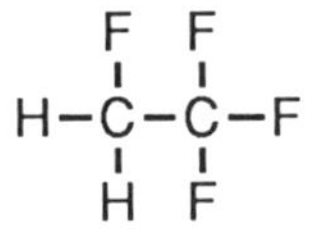

$$H-CHF-CF_3$$

Malathion
[(dimethoxyphosphinothioyl)thio]butanedioc acid diethyl ester

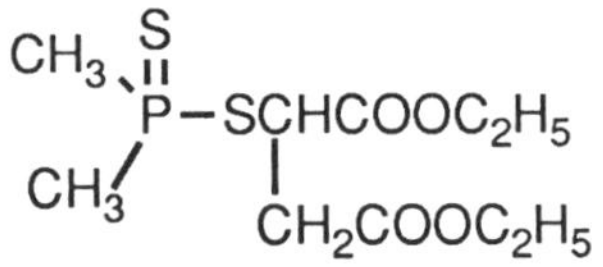

$$(CH_3)_2P(=S)-SCH(COOC_2H_5)CH_2COOC_2H_5$$

Methyl Mercury
methyl mercury chloride

$$CH_3Hg^+Cl^-$$

dimethyl mercury

$$CH_3-Hg-CH_3$$

Nicotine
3-(1-methyl-2-pyrrolidinyl)pyridine

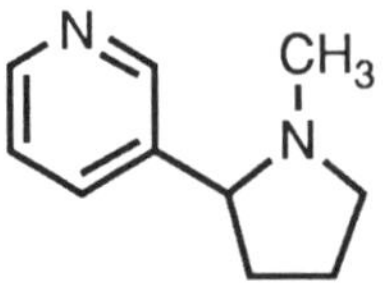

NTA (nitrilotriacetic acid)
N,N-bis(carboxymethyl)glycine

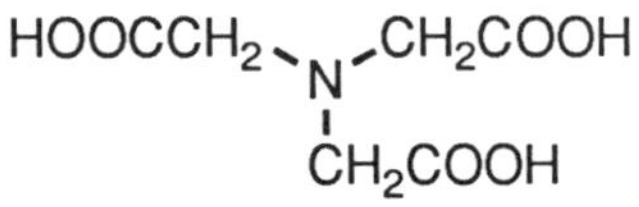

$$HOOCCH_2-N(CH_2COOH)-CH_2COOH$$

PAHs (polyaromatic hydrocarbons)
benzo(*a*)pyrene

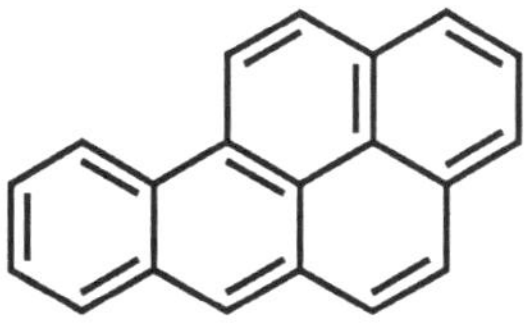

benzo[*b*]fluoranthene

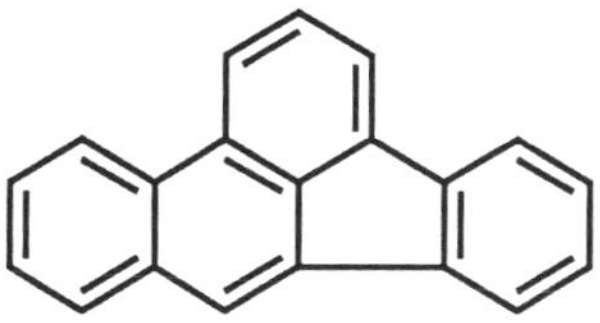

chrysene

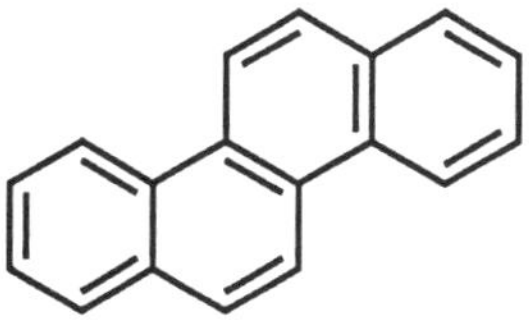

PAN
peroxyacetylnitrate

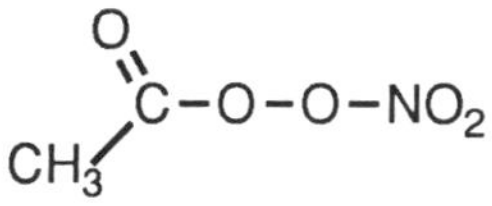

$CH_3C(=O)-O-O-NO_2$

Parathion
phosphorothioc O,O-diethyl O-(4-nitrophenyl) ester

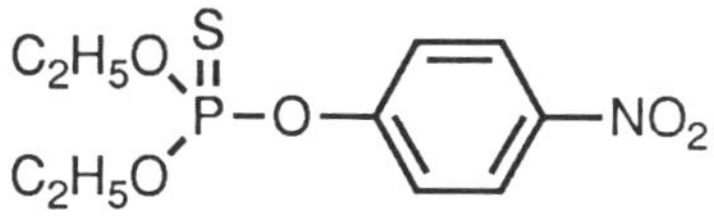

$(C_2H_5O)_2P(=S)-O-C_6H_4-NO_2$

PCBs
polychlorinated biphenyls

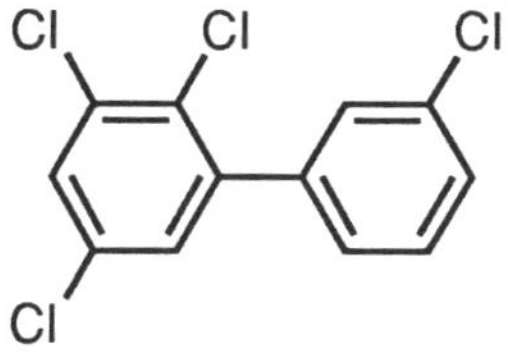

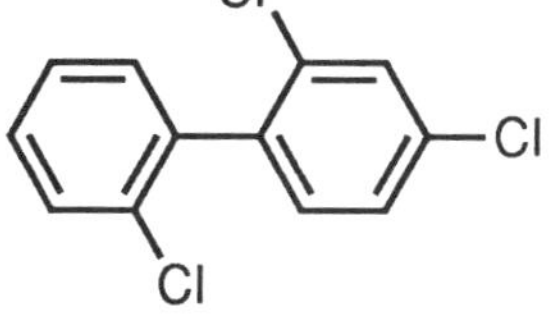

PCDFs
polychlorinated dibenzofurans

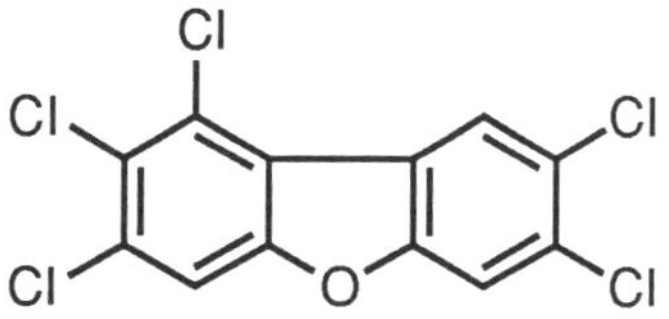

PCP
pentachlorophenol

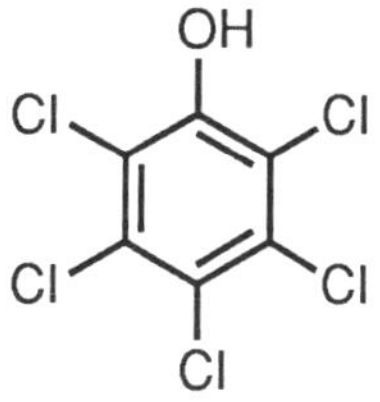

Phosphate
tripolyphosphate ion

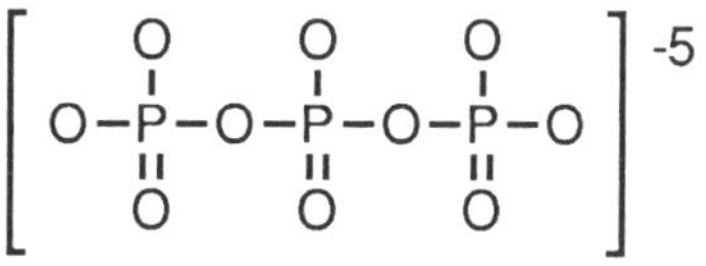

$$\left[O-\overset{O}{\overset{|}{P}}-O-\overset{O}{\overset{|}{P}}-O-\overset{O}{\overset{|}{P}}-O \right]^{-5}$$

phosphate ion

$$\left[O-\overset{O}{\overset{|}{P}}-O \right]^{-3}$$

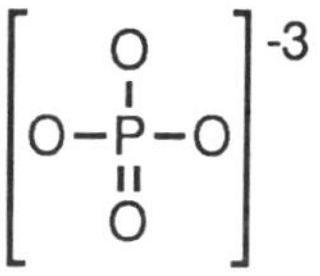

Phosvel
phenylphonothioic acid O-(4-bromo-2,5-dichlorophenyl) O-methyl ester

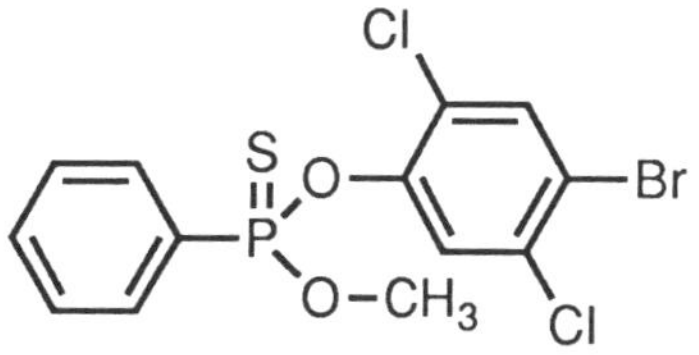

PVC
polyvinylchloride

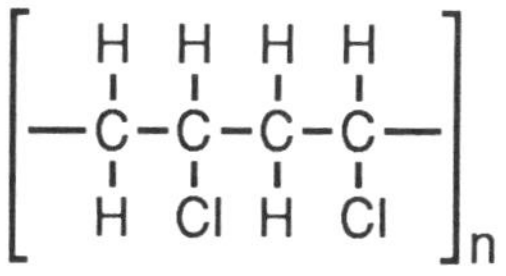

$$\left[-CH_2-CHCl-CH_2-CHCl- \right]_n$$

Pyrethrum
dalmatian insect powder composed of pyrethin (shown below), pyretol, pyrethotoxic acid, pyrethrosin, chrysanthemine and crysathemumic acid

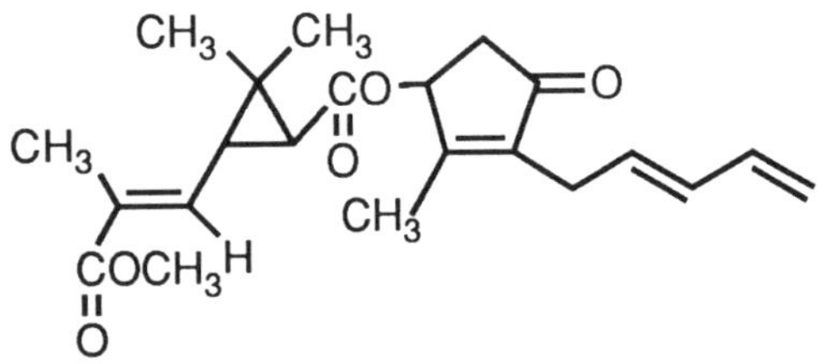

Strychnine

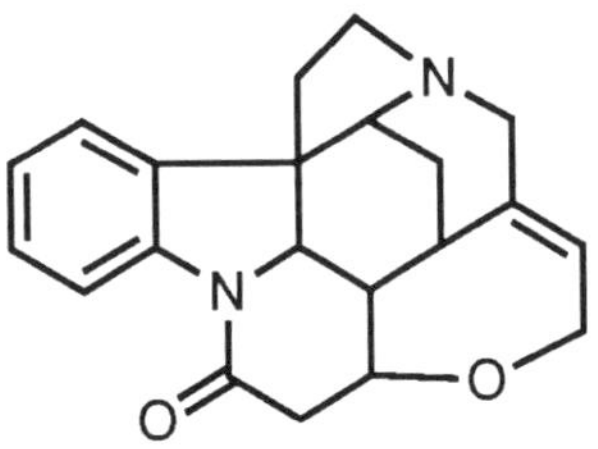

TCDD
2,3,7,8-tetrachlorodibenzoparadioxin

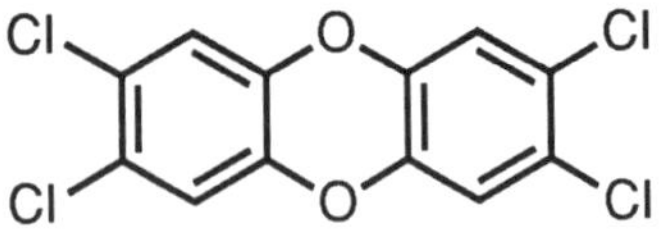

TCP
tricholrophenol

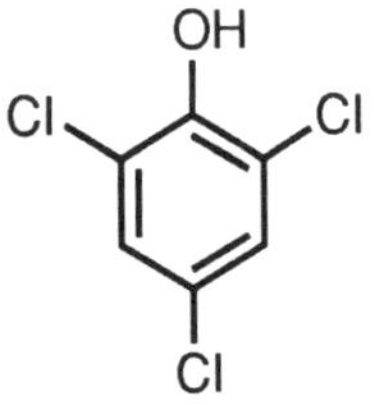

Toxaphene
a complex mixture of chlorinated camphenes, one example is shown

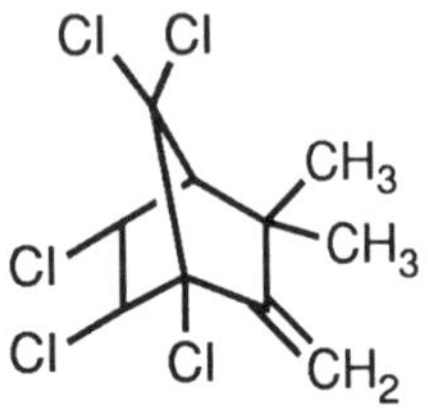

2,4-D
2,4-dichlorophenoxyacetic acid

$O-CH_2-COOH$

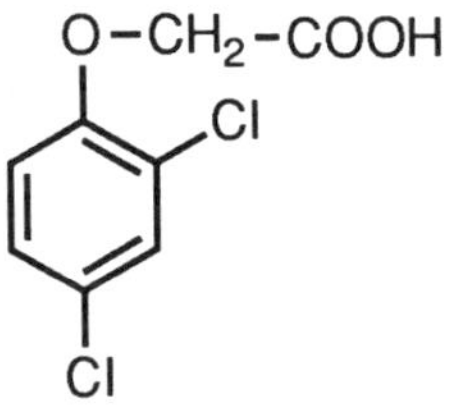

2,4,5-T
2,4,5-trichlorophenoxyacetic acid

$O-CH_2-COOH$

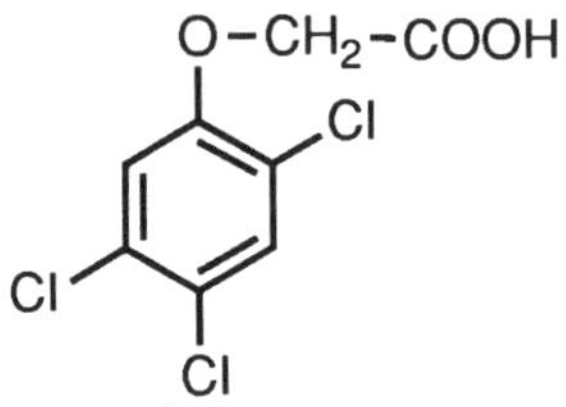

UFFI
urea polymer with formaldehyde

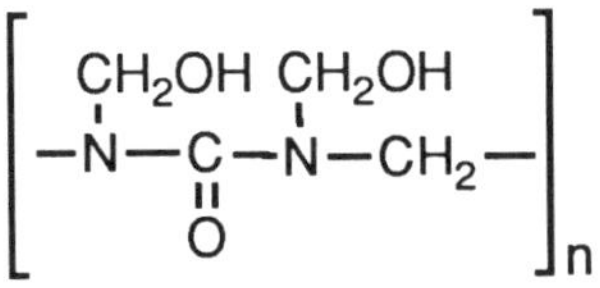

$[-N(CH_2OH)-C(=O)-N(CH_2OH)-CH_2-]_n$

VOCs (volatile organic compounds)
aliphatic hydrocarbons, eg octane

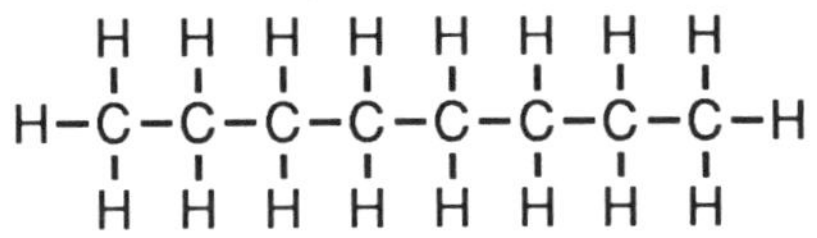

$H-(CH_2)_8-H$

aromatic hydrocarbons eg benzene

alcohols eg ethanol

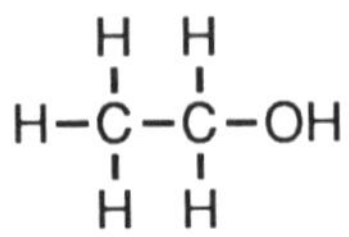

$H-CH_2-CH_2-OH$

INDEX

A book giving answers in plain English to the questions:

- What chemicals are entering the environment?
- Who is using them, and why?
- How hazardous are they to our health and the ecosystem?
- What steps are being taken to control them?

THE JOHN RYLANDS UNIVERSITY LIBRARY
The John Rylands Library